# Nuclear Engineering

# Nuclear Engineering

**Edited by**
Lindsay Garfield

Willford Press
www.willfordpress.com

Published by Willford Press,
118-35 Queens Blvd., Suite 400,
Forest Hills, NY 11375, USA



ISBN: 978-1-68285-485-3

**Cataloging-in-Publication Data**

Nuclear engineering / edited by Lindsay Garfield.
p. cm.
Includes bibliographical references and index.
ISBN 978-1-68285-485-3
1. Nuclear engineering. 2. Nuclear energy. I. Garfield, Lindsay.

TK9145 .N83 2018
621.48--dc23

For information on all Willford Press publications
visit our website at www.willfordpress.com

WILLFORD PRESS

Printed in China.

# Contents

# Preface

As the world is growing at a very rapid pace, so is the technology. Nuclear science and engineering refers to one such technology, which studies the characteristics and properties of the atomic nuclei. It has applications in the fields like geology, nuclear medicine, ion implantation, magnetic resonance imaging, etc. The topics covered under this subject are nuclear fusion, nuclear decay, nuclear fission, etc. This book is a compilation of chapters that discuss the most vital concepts in the field of nuclear engineering. It studies, analyses and upholds the pillars of this field and its utmost significance in modern times. The various sub-fields of nuclear science along with technological progress that have future implications are glanced at in this textbook. It aims to serve as a resource guide for students and experts alike and contribute to the growth of the discipline.

Given below is the chapter wise description of the book:

Chapter 1- Nuclear physics is a branch of physics that studies the atomic nucleus and its components. Nucleus is the central part of an atom comprising of protons and neutrons. It carries the whole mass of an atom. While protons are positively charged, neutrons have no charge. This is an introductory chapter which will introduce briefly all the significant aspects of atomic nucleus.

Chapter 2- Understanding the structure of the nucleus is central to nuclear physics. To understand nuclear structure, various models are applied, out of which the liquid drop model is the simplest that has proven to best comprehend the structure of a nucleus. However, some features of nucleus structure cannot be understood using this method for which other models like nuclear shell model, interacting Boson model, etc. are used. This section discusses in detail the theories and methodologies related to models of the nucleus.

Chapter 3- To detect radiation of an electron, it is necessary to understand the means by which an electron is enhanced. There are natural ways, however, artificial means are also used as radiation detection holds a significant place in science. A renowned artificial means is the particle accelerator, which accelerates the speed of charged particles to light speed and then contains them in beams. This chapter has been carefully written to provide an easy understanding of the varied facets of particle accelerators.

Chapter 4- Nuclear detectors use the energy loss of energetic materials to gather information about particles. A few types of nuclear detectors are gas detectors, solid state detectors, scintillation detectors, etc. The section strategically encompasses and incorporates the major types of nuclear detectors and their applications, providing a complete understanding.

Chapter 5- Nuclear reaction is a reaction in which two or more nuclei or their components collide to form a nuclide, which attributes different characteristics. Also it is necessary to know that there is output of energy and not reverse effect. For this, the energy gain or loss is calculated which is called Q-value. The more the Q-value, the faster the reaction takes place. The diverse applications of nuclear reactions and reactors in the current scenario have been thoroughly discussed in this chapter.

Chapter 6- The process by which a nucleus can emit particles such as alpha, beta and gamma without being subjected to any external disturbance is called radioactivity. The process is governed by radioactive decay in which a nucleus disintegrates when it emits charged particles. In order to completely understand radioactivity, it is necessary to understand the processes related to it. The following section elucidates the varied processes and mechanisms associated with this area of study.

Chapter 7- It becomes difficult for scientists to analyze the composition of materials when the particle's presence extremely low. This calls for techniques to be used for quantifying and identifying materials. A few of the techniques used in this process are neutron activation analysis, Rutherford backscattering spectrometry, nuclear reaction analysis, etc. The chapter serves as a source to understand the major categories related to nuclear techniques.

Indeed, my job was extremely crucial and challenging as I had to ensure that every chapter is informative and structured in a student-friendly manner. I am thankful for the support provided by my family and colleagues during the completion of this book.

**Editor**

1

# Fundamentals of Nuclear Physics

Nuclear physics is a branch of physics that studies the atomic nucleus and its components. Nucleus is the central part of an atom comprising of protons and neutrons. It carries the whole mass of an atom. While protons are positively charged, neutrons have no charge. This is an introductory chapter which will introduce briefly all the significant aspects of atomic nucleus.

## Nuclear Physics

Nuclear physics is the field of physics that studies atomic nuclei and their constituents and interactions. Other forms of nuclear matter are also studied. Nuclear physics should not be confused with atomic physics, which studies the atom as a whole, including its electrons.

Discoveries in nuclear physics have led to applications in many fields. This includes nuclear power, nuclear weapons, nuclear medicine and magnetic resonance imaging, industrial and agricultural isotopes, ion implantation in materials engineering, and radiocarbon dating in geology and archaeology. Such applications are studied in the field of nuclear engineering.

Particle physics evolved out of nuclear physics and the two fields are typically taught in close association. Nuclear astrophysics, the application of nuclear physics to astrophysics, is crucial in explaining the inner workings of stars and the origin of the chemical elements.

### History

Henri Becquerel

The history of nuclear physics as a discipline distinct from atomic physics starts with the discovery of radioactivity by Henri Becquerel in 1896, while investigating phosphorescence in uranium salts. The discovery of the electron by J. J. Thomson a year later was an indication that the atom had internal structure. At the beginning of the 20th century the accepted model of the atom was J. J. Thomson's "plum pudding" model in which the atom was a positively charged ball with smaller negatively charged electrons embedded inside it.

In the years that followed, radioactivity was extensively investigated, notably by Marie and Pierre Curie as well as by Ernest Rutherford and his collaborators. By the turn of the century physicists had also discovered three types of radiation emanating from atoms, which they named alpha, beta, and gamma radiation. Experiments by Otto Hahn in 1911 and by James Chadwick in 1914 discovered that the beta decay spectrum was continuous rather than discrete. That is, electrons were ejected from the atom with a continuous range of energies, rather than the discrete amounts of energy that were observed in gamma and alpha decays. This was a problem for nuclear physics at the time, because it seemed to indicate that energy was not conserved in these decays.

The 1903 Nobel Prize in Physics was awarded jointly to Becquerel for his discovery and to Marie and Pierre Curie for their subsequent research into radioactivity. Rutherford was awarded the Nobel Prize in Chemistry in 1908 for his "investigations into the disintegration of the elements and the chemistry of radioactive substances".

In 1905 Albert Einstein formulated the idea of mass–energy equivalence. While the work on radioactivity by Becquerel and Marie Curie predates this, an explanation of the source of the energy of radioactivity would have to wait for the discovery that the nucleus itself was composed of smaller constituents, the nucleons.

## Rutherford's Team Discovers the Nucleus

In 1906 Ernest Rutherford published "Retardation of the α Particle from Radium in passing through matter." Hans Geiger expanded on this work in a communication to the Royal Society with experiments he and Rutherford had done, passing alpha particles through air, aluminum foil and gold leaf. More work was published in 1909 by Geiger and Ernest Marsden, and further greatly expanded work was published in 1910 by Geiger. In 1911–1912 Rutherford went before the Royal Society to explain the experiments and propound the new theory of the atomic nucleus as we now understand it.

The key experiment behind this announcement was performed in 1910 at the University of Manchester: Ernest Rutherford's team performed a remarkable experiment in which Geiger and Marsden under Rutherford's supervision fired alpha particles (helium nuclei) at a thin film of gold foil. The plum pudding model had predicted that the alpha particles should come out of the foil with their trajectories being at most slightly bent. But Rutherford instructed his team to look for something that shocked him to observe: a few particles were scattered through large angles, even completely backwards in some cases. He likened it to firing a bullet at tissue paper and having it bounce off. The discovery, with Rutherford's analysis of the data in 1911, led to the Rutherford model of the atom, in which the atom had a very small, very dense nucleus containing most of its mass, and consisting of heavy positively charged particles with embedded electrons in order to balance out the charge (since the neutron was unknown). As an example, in this model (which is not the modern one) nitrogen-14 consisted of a nucleus with 14 protons and 7 electrons (21 total particles) and the nucleus was surrounded by 7 more orbiting electrons.

The Rutherford model worked quite well until studies of nuclear spin were carried out by Franco Rasetti at the California Institute of Technology in 1929. By 1925 it was known that protons and electrons each had a spin of $\frac{1}{2}$. In the Rutherford model of nitrogen-14, 20 of the total 21 nuclear particles should have paired up to cancel each other's spin, and the final odd particle should have left the nucleus with a net spin of $\frac{1}{2}$. Rasetti discovered, however, that nitrogen-14 had a spin of 1.

## James Chadwick Discovers the Neutron

In 1932 Chadwick realized that radiation that had been observed by Walther Bothe, Herbert Becker, Irène and Frédéric Joliot-Curie was actually due to a neutral particle of about the same mass as the proton, that he called the neutron (following a suggestion from Rutherford about the need for such a particle). In the same year Dmitri Ivanenko suggested that there were no electrons in the nucleus — only protons and neutrons — and that neutrons were spin $\frac{1}{2}$ particles which explained the mass not due to protons. The neutron spin immediately solved the problem of the spin of nitrogen-14, as the one unpaired proton and one unpaired neutron in this model each contributed a spin of $\frac{1}{2}$ in the same direction, giving a final total spin of 1.

With the discovery of the neutron, scientists could at last calculate what fraction of binding energy each nucleus had, by comparing the nuclear mass with that of the protons and neutrons which composed it. Differences between nuclear masses were calculated in this way. When nuclear reactions were measured, these were found to agree with Einstein's calculation of the equivalence of mass and energy to within 1% as of 1934.

## Proca's Equations of the Massive Vector Boson Field

Alexandru Proca was the first to develop and report the massive vector boson field equations and a theory of the mesonic field of nuclear forces. Proca's equations were known to Wolfgang Pauli who mentioned the equations in his Nobel address, and they were also known to Yukawa, Wentzel, Taketani, Sakata, Kemmer, Heitler, and Fröhlich who appreciated the content of Proca's equations for developing a theory of the atomic nuclei in Nuclear Physics.

## Yukawa's Meson Postulated to Bind Nuclei

In 1935 Hideki Yukawa proposed the first significant theory of the strong force to explain how the nucleus holds together. In the Yukawa interaction a virtual particle, later called a meson, mediated a force between all nucleons, including protons and neutrons. This force explained why nuclei did not disintegrate under the influence of proton repulsion, and it also gave an explanation of why the attractive strong force had a more limited range than the electromagnetic repulsion between protons. Later, the discovery of the pi meson showed it to have the properties of Yukawa's particle.

With Yukawa's papers, the modern model of the atom was complete. The center of the atom contains a tight ball of neutrons and protons, which is held together by the strong nuclear force, unless it is too large. Unstable nuclei may undergo alpha decay, in which they emit an energetic helium nucleus, or beta decay, in which they eject an electron (or positron). After one of these decays the resultant nucleus may be left in an excited state, and in this case it decays to its ground state by emitting high energy photons (gamma decay).

The study of the strong and weak nuclear forces (the latter explained by Enrico Fermi via Fermi's interaction in 1934) led physicists to collide nuclei and electrons at ever higher energies. This research became the science of particle physics, the crown jewel of which is the standard model of particle physics which describes the strong, weak, and electromagnetic forces.

## Modern Nuclear Physics

A heavy nucleus can contain hundreds of nucleons. This means that with some approximation it can be treated as a classical system, rather than a quantum-mechanical one. In the resulting liquid-drop model, the nucleus has an energy which arises partly from surface tension and partly from electrical repulsion of the protons. The liquid-drop model is able to reproduce many features of nuclei, including the general trend of binding energy with respect to mass number, as well as the phenomenon of nuclear fission.

Superimposed on this classical picture, however, are quantum-mechanical effects, which can be described using the nuclear shell model, developed in large part by Maria Goeppert Mayer and J. Hans D. Jensen. Nuclei with certain numbers of neutrons and protons (the magic numbers 2, 8, 20, 28, 50, 82, 126, ...) are particularly stable, because their shells are filled.

Other more complicated models for the nucleus have also been proposed, such as the interacting boson model, in which pairs of neutrons and protons interact as bosons, analogously to Cooper pairs of electrons.

Ab initio methods try to solve the nuclear many-body problem from the ground up, starting from the nucleons and their interactions.

Much of current research in nuclear physics relates to the study of nuclei under extreme conditions such as high spin and excitation energy. Nuclei may also have extreme shapes (similar to that of Rugby balls or even pears) or extreme neutron-to-proton ratios. Experimenters can create such nuclei using artificially induced fusion or nucleon transfer reactions, employing ion beams from an accelerator. Beams with even higher energies can be used to create nuclei at very high temperatures, and there are signs that these experiments have produced a phase transition from normal nuclear matter to a new state, the quark–gluon plasma, in which the quarks mingle with one another, rather than being segregated in triplets as they are in neutrons and protons.

## Nuclear Decay

Eighty elements have at least one stable isotope which is never observed to decay, amounting to a total of about 254 stable isotopes. However, thousands of isotopes have been characterized as unstable. These "radioisotopes" decay over time scales ranging from fractions of a second to trillions of years. Plotted on a chart as a function of atomic and neutron numbers, the binding energy of the nuclides forms what is known as the valley of stability. Stable nuclides lie along the bottom of this energy valley, while increasingly unstable nuclides lie up the valley walls, that is, have weaker binding energy.

The most stable nuclei fall within certain ranges or balances of composition of neutrons and protons: too few or too many neutrons (in relation to the number of protons) will cause it to decay. For example, in beta decay a nitrogen-16 atom (7 protons, 9 neutrons) is converted to an oxygen-16 atom (8 protons, 8 neutrons) within a few seconds of being created. In this decay a neutron in the nitrogen nucleus is converted by the weak interaction into a proton, an electron and an antineutrino. The element is transmuted to another element, with a different number of protons.

In alpha decay (which typically occurs in the heaviest nuclei) the radioactive element decays by emitting a helium nucleus (2 protons and 2 neutrons), giving another element, plus helium-4. In many cases this process continues through several steps of this kind, including other types of decays (usually beta decay) until a stable element is formed.

In gamma decay, a nucleus decays from an excited state into a lower energy state, by emitting a gamma ray. The element is not changed to another element in the process (no nuclear transmutation is involved).

Other more exotic decays are possible. For example, in internal conversion decay, the energy from an excited nucleus may eject one of the inner orbital electrons from the atom, in a process which produces high speed electrons, but is not beta decay, and (unlike beta decay) does not transmute one element to another.

## Nuclear Fusion

In nuclear fusion, two low mass nuclei come into very close contact with each other, so that the strong force fuses them. It requires a large amount of energy for the strong or nuclear forces to overcome the electrical repulsion between the nuclei in order to fuse them; therefore nuclear fusion can only take place at very high temperatures or high pressures. When nuclei fuse, a very large amount of energy is released and the combined nucleus assumes a lower energy level. The binding energy per nucleon increases with mass number up to nickel-62. Stars like the Sun are powered by the fusion of four protons into a helium nucleus, two positrons, and two neutrinos. The uncontrolled fusion of hydrogen into helium is known as thermonuclear runaway. A frontier in current research at various institutions, for example the Joint European Torus (JET) and ITER, is the development of an economically viable method of using energy from a controlled fusion reaction. Nuclear fusion is the origin of the energy (including in the form of light and other electromagnetic radiation) produced by the core of all stars including our own Sun.

## Nuclear Fission

Nuclear fission is the reverse process to fusion. For nuclei heavier than nickel-62 the binding energy per nucleon decreases with the mass number. It is therefore possible for energy to be released if a heavy nucleus breaks apart into two lighter ones.

The process of alpha decay is in essence a special type of spontaneous nuclear fission. It is a highly asymmetrical fission because the four particles which make up the alpha particle are especially tightly bound to each other, making production of this nucleus in fission particularly likely.

From certain of the heaviest nuclei whose fission produces free neutrons, and which also easily absorb neutrons to initiate fission, a self-igniting type of neutron-initiated fission can be obtained, in a chain reaction. Chain reactions were known in chemistry before physics, and in fact many familiar processes like fires and chemical explosions are chemical chain reactions. The fission or "nuclear" chain-reaction, using fission-produced neutrons, is the source of energy for nuclear power plants and fission type nuclear bombs, such as those detonated in Hiroshima and Nagasaki, Japan, at the end of World War II. Heavy nuclei such as uranium and thorium may also undergo spontaneous fission, but they are much more likely to undergo decay by alpha decay.

For a neutron-initiated chain reaction to occur, there must be a critical mass of the relevant isotope present in a certain space under certain conditions. The conditions for the smallest critical mass require the conservation of the emitted neutrons and also their slowing or moderation so that there is a greater cross-section or probability of them initiating another fission. In two regions of Oklo, Gabon, Africa, natural nuclear fission reactors were active over 1.5 billion years ago. Measurements of natural neutrino emission have demonstrated that around half of the heat emanating from the Earth's core results from radioactive decay. However, it is not known if any of this results from fission chain reactions.

## Production of "Heavy" Elements (Atomic Number Greater Than Five)

According to the theory, as the Universe cooled after the Big Bang it eventually became possible for common subatomic particles as we know them (neutrons, protons and electrons) to exist. The most common particles created in the Big Bang which are still easily observable to us today were protons and electrons (in equal numbers). The protons would eventually form hydrogen atoms. Almost all the neutrons created in the Big Bang were absorbed into helium-4 in the first three minutes after the Big Bang, and this helium accounts for most of the helium in the universe today.

Some relatively small quantities of elements beyond helium (lithium, beryllium, and perhaps some boron) were created in the Big Bang, as the protons and neutrons collided with each other, but all of the "heavier elements" (carbon, element number 6, and elements of greater atomic number) that we see today, were created inside stars during a series of fusion stages, such as the proton-proton chain, the CNO cycle and the triple-alpha process. Progressively heavier elements are created during the evolution of a star.

Since the binding energy per nucleon peaks around iron (56 nucleons), energy is only released in fusion processes involving smaller atoms than that. Since the creation of heavier nuclei by fusion requires energy, nature resorts to the process of neutron capture. Neutrons (due to their lack of charge) are readily absorbed by a nucleus. The heavy elements are created by either a *slow* neutron capture process (the so-called *s* process) or the *rapid*, or *r* process. The *s* process occurs in thermally pulsing stars (called AGB, or asymptotic giant branch stars) and takes hundreds to thousands of years to reach the heaviest elements of lead and bismuth. The *r* process is thought to occur in supernova explosions which provide the necessary conditions of high temperature, high neutron flux and ejected matter. These stellar conditions make the successive neutron captures very fast, involving very neutron-rich species which then beta-decay to heavier elements, especially at the so-called waiting points that correspond to more stable nuclides with closed neutron shells (magic numbers).

## Nuclear Mass

Nucleus contains almost entire mass of an atom. The mass of a proton is ~ 1.6725 x 10 $^{-27}$ kg or 938.26 MeV and that of a neutron is 1.6748 x 10 $^{-27}$ kg, i.e. 939.55 MeV. The atomic masses denoted as atomic mass unit (a.m.u.) correspond to 1/12$^{th}$ of the mass of $^{12}$C atom. Its value is 1.660566 x 10 $^{-27}$ kg = 931.5016 ?± 0.0026 MeV/c$^2$. * Electrons, the other constituents of atoms are much lighter (1/1840 of a proton mass) and hence negligible contribution to atomic mass.

*As per Eienstein's Relativity theory, mass and energy are interconvertible by an equation E = mc$^2$. Therefore, amu can be converted to MeV.

## Nuclear Dimension

Comparing with the size of an atom, nucleus is 1000000 times smaller. You might be surprised how one can determine any object of such small dimension? In fact no microscope exists till date to view such small dimensional object. However, Physicists have other way out to determine such small physical structure. From elementary quantum mechanics it is known that every particle has a wave associated with it, which is known as matter wave or de Broglie wave. The de Broglie wavelength* of 100 mega electron volt (MeV) electron is of similar order of nuclear dimension and hence this can be used as a probe to determine the size of a nucleus. Moreover, the interaction between electron and nucleus is very well known, i.e. electromagnetic and hence electron is chosen as a probe to get the idea about the dimension of a nucleus. Let us see how one gets the idea about nuclear dimension using experimental set up depicted in the figure:

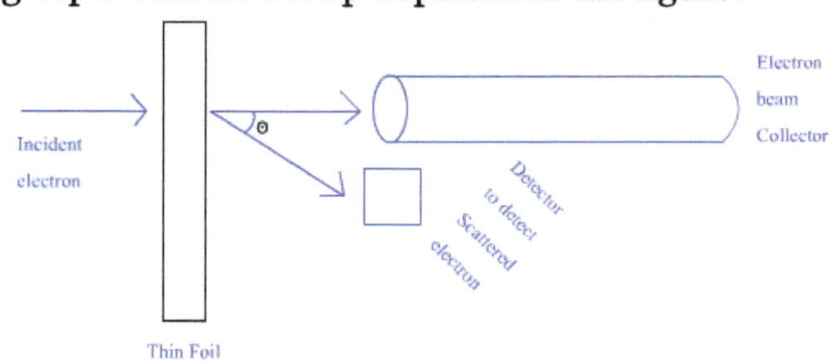


Schematic of experimental set up to measure angular deflection of electron from nuclei, where q is the scattering angle. Detector is placed at different q to detect the scattered electrons. A detector (Faraday Cup for precise measurement of charge) is placed at the back of the target to collect the incident beam of electrons.

A high-energy electron beam is bombarded on a thin material target under study and the probability of various angular deflections is observed. A typical result of this experiment is shown in the figure, which basically shows the angular distribution of electrons scattered from various nuclei.

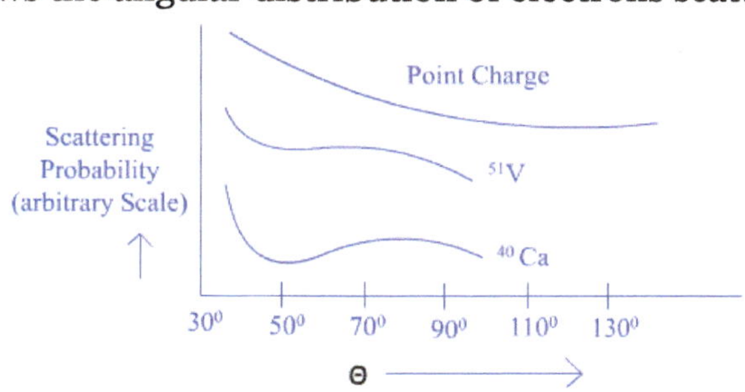


Angular distribution of electrons scattered from various nuclei. Relative intensity is shown in log scale as a function of scattering angle.

Now we try to understand how this distribution helps us to obtain the dimension of a nucleus.

Let us assume that ρ(r) is some density distribution for the nucleons in the nucleus, and it is same for neutrons and protons. Let ρ(o) be the nucleon density near the centre of the nucleus, R the radius at which the density has decreased by a factor 2 below the central value, and a is a measure of the surface thickness such that the distance over which the density falls from 90% of ρ(o) to 10% of ρ(o) is 4.4a. Let ρ(r) is related to all these physical parameters by the relation

$$\rho(r) = \frac{\rho(0)}{1+\exp\left[(r-R)/a\right]}$$

If we plot this function, it will look like the one as shown in the figure below:

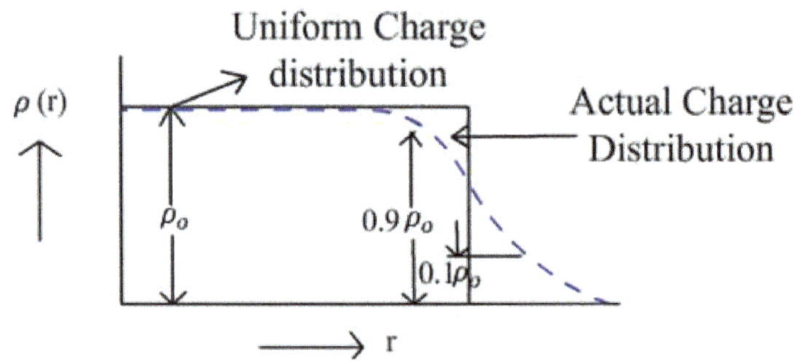


Plot of equation to show the variation of r (r) with r.

The scattering cross-section# can be calculated using quantum mechanical scattering formalism. Using ρ(r) as a fitting parameter, the angular deflection of electrons can be calculated and fitted with the plots of figure. The value of ρ(r) is chosen for the best fitted curve and the following information is obtained.

ρ (o) ≅ 1.65 x 1044 nucleons/m$^3$ = 0.165 nucleons/F$^3$.

R ≅ 1.07 A$^{1/3}$ F

a ≅ 0.55 F

where F stands for 'Fermi' and is equal to $10^{-15}$ m, as the unit of length. The above analysis gives the idea of nuclear dimension, which is almost $10^6$ times less than the atomic dimension.

Now just imagine that almost entire mass of an atom is located in such small dimension and therefore the mass density of a nucleus is extremely high. Physical properties of this fascinating object are therefore extremely important from both fundamental understanding and engineering applications point of views.

*Every particle has an associated wave with it, known as matter wave or de Broglie wave. The wavelength (λ) of this wave = h/p, where h is the Planck's constant and p is the momentum of the particle. If you calculate λ value of 100 MeV electron, you will find its wavelength is of the order of nuclear dimension. This is the reason for choosing this as a probe for this experiment.

The cross section of any scattering process basically signifies the probability of occurrence of that process. In this case the angular distribution of electrons gives an idea what is the probability of electrons scattered at different angles. For more information about cross section.

For quantum mechanical scattering you have to go through a quantum mechanics course which includes this topic like partial wave analysis, Born's approximation etc.

## Neutron Magnetic Moment

The neutron magnetic moment is the intrinsic magnetic dipole moment of the neutron, symbol $\mu_n$. Protons and neutrons, both nucleons, comprise the nucleus of atoms, and both nucleons behave as small magnets whose strengths are measured by their magnetic moments. The neutron interacts with normal matter primarily through the nuclear force and through its magnetic moment. The neutron's magnetic moment is exploited to probe the atomic structure of materials using scattering methods and to manipulate the properties of neutron beams in particle accelerators. The neutron was determined to have a magnetic moment by indirect methods in the mid 1930s. Luis Alvarez and Felix Bloch made the first accurate, direct measurement of the neutron's magnetic moment in 1940. The existence of the neutron's magnetic moment indicates the neutron is not an elementary particle. For an elementary particle to have an intrinsic magnetic moment, it must have both spin and electric charge. The neutron has spin 1/2 $\hbar$, but it has no net charge. The existence of the neutron's magnetic moment was puzzling and defied a correct explanation until the quark model for particles was developed in the 1960s. The neutron is composed of three quarks, and the magnetic moments of these elementary particles combine to give the neutron its magnetic moment.

### Description

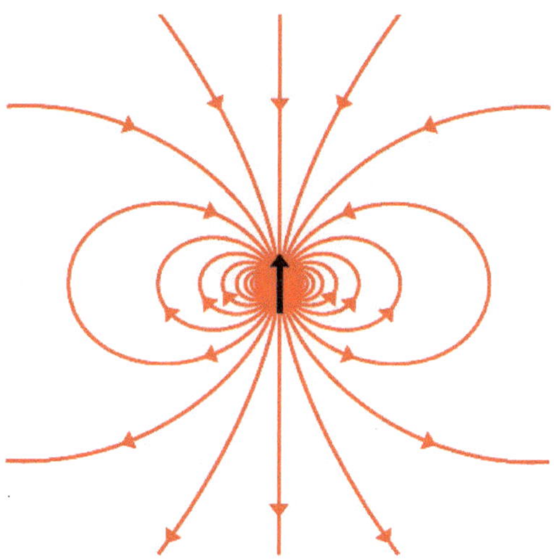

Schematic diagram depicting the spin of the neutron as the black arrow and magnetic field lines associated with the neutron's negative magnetic moment. The spin of the neutron is upward in this diagram, but the magnetic field lines at the center of the dipole are downward.

The best available measurement for the value of the magnetic moment of the neutron is $\mu_n$ = −1.91304272(45) $\mu_N$. Here $\mu_N$ is the nuclear magneton, a physical constant and standard unit for the magnetic moments of nuclear components. In SI units, $\mu_n$ = −9.6623647(23)×10$^{-27}$ J·T$^{-1}$. A magnetic moment is a vector quantity, and the direction of the neutron's magnetic moment is defined by its spin. The torque on the neutron resulting from an external magnetic field is towards aligning the neutron's spin vector opposite to the magnetic field vector.

The nuclear magneton is the spin magnetic moment of a Dirac particle, a charged, spin 1/2 elementary particle, with a proton's mass $m_p$. In SI units, the nuclear magneton is

$$\mu_N = \frac{e\hbar}{2m_p},$$

where $e$ is the elementary charge and $\hbar$ is the reduced Planck constant. The magnetic moment of this particle is parallel to its spin. Since the neutron has no charge, it should have no magnetic moment by this expression. The non-zero magnetic moment of the neutron indicates that it is not an elementary particle. The sign of the neutron's magnetic moment is that of a negatively charged particle. Similarly, the fact that the magnetic moment of the proton, $\mu_p = 2.793\ \mu_N$, is not equal to 1 $\mu_N$ indicates that it too is not an elementary particle. Protons and neutrons are composed of quarks, and the magnetic moments of the quarks can be used to compute the magnetic moments of the nucleons.

Although the neutron interacts with normal matter primarily through either nuclear or magnetic forces, the magnetic interactions are about seven orders of magnitude weaker than the nuclear interactions. The influence of the neutron's magnetic moment is therefore only apparent for low energy, or slow, neutrons. Because the value for the magnetic moment is inversely proportional to particle mass, the nuclear magneton is about 1/2000 as large as the Bohr magneton. The magnetic moment of the electron is therefore about 1000 times larger than that of the neutron.

The magnetic moment of the antineutron has the same magnitude as, but has the opposite sign, that of the neutron.

## Measurement

Soon after the neutron was discovered in 1932, indirect evidence suggested the neutron had an unexpected non-zero value for its magnetic moment. Attempts to measure the neutron's magnetic moment originated with the discovery by Otto Stern in 1933 in Hamburg that the proton had an anomalously large magnetic moment. The proton's magnetic moment had been determined by measuring the deflection of a beam of molecular hydrogen by a magnetic field. Stern won the Nobel Prize in 1943 for this discovery.

By 1934 groups led by Stern, now in Pittsburgh, and I. I. Rabi in New York had independently measured the magnetic moments of the proton and deuteron. The measured values for these particles were only in rough agreement between the groups, but the Rabi group confirmed the earlier Stern measurements that the magnetic moment for the proton was unexpectedly large. Since a deuteron is composed of a proton and a neutron with aligned spins, the neutron's magnetic moment could be inferred by subtracting the deuteron and proton magnetic moments. The resulting value was not zero and had sign opposite to that of the proton. Values for the magnetic moment of the neutron were also determined by R. Bacher at Ann Arbor (1933) and I.Y. Tamm and S.A. Altshuler in the Soviet Union (1934) from studies of the hyperfine structure of atomic spectra. Although Tamm and Altshuler's estimate had the correct sign and order of magnitude ($\mu_n = -0.5\ \mu_N$), the result was met with skepticism. By the late 1930s, accurate values for the magnetic moment of the neutron had been deduced by the Rabi group using measurements employing newly developed nuclear magnetic resonance techniques. The large value for the proton's magnetic moment and the inferred negative value for the neutron's magnetic moment were unexpected and could not be explained. The anomalous values for the magnetic moments of the nucleons would remain a puzzle until the quark model was developed in the 1960s.

The refinement and evolution of the Rabi measurements led to the discovery in 1939 that the deuteron also possessed an electric quadrupole moment. This electrical property of the deuteron had been interfering with the measurements by the Rabi group. The discovery meant that the physical shape of the deuteron was not symmetric, which provided valuable insight into the nature of the nuclear force binding nucleons. Rabi was awarded the Nobel Prize in 1944 for his resonance method for recording the magnetic properties of atomic nuclei.

The value for the neutron's magnetic moment was first directly measured by Luis Alvarez and Felix Bloch at Berkeley, California in 1940, using an extension of the magnetic resonance methods developed by Rabi. Alvarez and Bloch determined the magnetic moment of the neutron to be $\mu_n = -1.93(2)\,\mu_N$. By directly measuring the magnetic moment of free neutrons, or individual neutrons free of the nucleus, Alvarez and Bloch resolved all doubts and ambiguities about this anomalous property of neutrons.

## Neutron *g*-factor and Gyromagnetic Ratio

The magnetic moment of a nucleon is sometimes expressed in terms of its *g*-factor, a dimensionless scalar. The convention defining the *g*-factor for composite particles, such as the neutron or proton, is

$$\boldsymbol{\mu} = \frac{g\mu_N}{\hbar}\mathbf{I}$$

where $\mu$ is the intrinsic magnetic moment, $I$ is the spin angular momentum, and $g$ is the effective *g*-factor. While the g-factor is dimensionless, for composite particles it is defined relative to the natural unit of the nuclear magneton. For the neutron, $I$ is $1/2\,\hbar$, so the neutron's *g*-factor, symbol $g_n$, is −3.82608545(90).

The gyromagnetic ratio, symbol $\gamma$, of a particle or system is the ratio of its magnetic moment to its spin angular momentum, or

$$\boldsymbol{\mu} = \gamma\mathbf{I}$$

For nucleons, the ratio is conventionally written in terms of the proton mass and charge, by the formula

$$\gamma = \frac{g\mu_N}{\hbar} = g\frac{e}{2m_p}$$

The neutron's gyromagnetic ratio, symbol $\gamma_n$, is $-1.83247179(43)\times10^8$ rad·s$^{-1}$·T$^{-1}$. The gyromagnetic ratio is also the ratio between the observed angular frequency of Larmor precession (in rad s$^{-1}$) and the strength of the magnetic field in nuclear magnetic resonance applications, such as in MRI imaging. For this reason, the value of $\gamma_n$ is often given in units of MHz/T. The quantity $\gamma_n/2\pi$ ("gamma bar") is therefore convenient, which has the value −29.1646943(69) MHz·T$^{-1}$.

## Physical Significance

Direction of Larmor precession for a neutron. The central arrow denotes the magnetic field, the small red arrow the spin of the neutron.

When a neutron is put into a magnetic field produced by an external source, it is subject to a torque tending to orient its magnetic moment parallel to the field (hence its spin antiparallel to the field). Like any magnet, the amount of this torque is proportional both to the magnetic moment and the external magnetic field. Since the neutron has spin angular momentum, this torque will cause the neutron to precess with a well-defined frequency, called the Larmor frequency. It is this phenomenon that enables the measurement of nuclear properties through nuclear magnetic resonance. The Larmor frequency can be determined by the product of the gyromagnetic ratio with the magnetic field strength. Since the sign of $\gamma_n$ is negative, the neutron's spin angular momentum precesses counterclockwise about the direction of the external magnetic field.

The interactions of the neutron's magnetic moment with an external magnetic field were exploited to finally determine the spin of the neutron. In 1949, Hughes and Burgy measured neutrons reflected from a ferromagnetic mirror and found that the angular distribution of the reflections was consistent with spin 1/2. In 1954, Sherwood, Stephenson, and Bernstein employed neutrons in a Stern-Gerlach experiment that used a magnetic field to separate the neutron spin states. They recorded the two such spin states, consistent with a spin 1/2 particle. Until these measurements, the possibility that the neutron was a spin 3/2 particle could not have been ruled out.

Since neutrons are neutral particles, they do not have to overcome Coulomb repulsion as they approach charged targets, as experienced by protons or alpha particles. Neutrons can deeply penetrate matter. The magnetic moment of the neutron has therefore been exploited to probe the properties of matter using scattering or diffraction techniques. These methods provide information that is complementary to X-ray spectroscopy. In particular, the magnetic moment of the neutron is used to determine magnetic properties of materials at length scales of 1–100 Å using cold or thermal neutrons. Bertram Brockhouse and Clifford Shull won the Nobel Prize in physics in 1994 for developing these scattering techniques.

Without an electric charge, neutron beams cannot be controlled by the conventional electromagnetic methods employed for particle accelerators. The magnetic moment of the neutron allows some control of neutrons using magnetic fields, however, including the formation of polarized neutron beams. One technique employs the fact that cold neutrons will reflect from some magnetic materials at great efficiency when scattered at small grazing angles. The reflection preferentially selects particular spin states, thus polarizing the neutrons. Neutron magnetic mirrors and guides use this total internal reflection phenomena to control beams of slow neutrons.

Since an atomic nucleus consists of a bound state of protons and neutrons, the magnetic moments of the nucleons contribute to the nuclear magnetic moment, or the magnetic moment for the nucleus as a whole. The nuclear magnetic moment also includes contributions from the orbital motion of the nucleons. The deuteron has the simplest example of a nuclear magnetic moment, with measured value 0.857 $\mu_N$. This value is within 3% of the sum of the moments of the proton and neutron, which gives 0.879 $\mu_N$. In this calculation, the spins of the nucleons are aligned, but their magnetic moments offset because of the neutron's negative magnetic moment.

## Anomalous Magnetic Moments and Meson Physics

The anomalous values for the magnetic moments of the nucleons presented a theoretical quandary for the 30 years from the time of their discovery in the early 1930s to the development of the quark model in the 1960s. Considerable theoretical efforts were expended in trying to understand the origins of these magnetic moments, but the failures of these theories were glaring. Much of the theoretical focus was on developing a nuclear-force equivalence to the remarkably successful theory explaining the small anomalous magnetic moment of the electron.

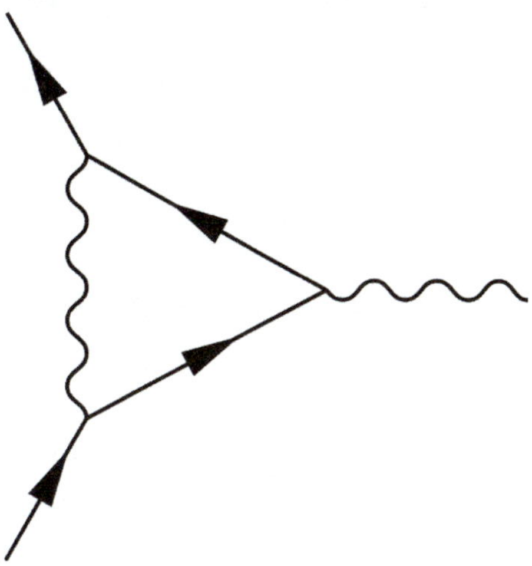

One-loop correction to a fermion's magnetic dipole moment. The solid lines at top and bottom represent the fermion (electron or nucleon), the wavey lines represent the photon mediating the electromagnetic force. The middle solid lines represent a virtual pair of particles, electron and positron for QED, pions for the nuclear force.

In quantum electrodynamics (QED), the anomalous magnetic moment of a particle stems from the small contributions of quantum mechanical fluctuations to the magnetic moment of that particle. The g-factor for a "Dirac" magnetic moment is predicted to be $g = -2$ for a negatively charged, spin 1/2 particle. For particles such as the electron, this "classical" result differs from the observed value by a small fraction of a percent; the difference compared to the classical value is the anomalous magnetic moment. The actual g-factor for the electron is measured to be −2.00231930436153(53).

QED results from the mediation of the electromagnetic force by photons. The physical picture is that the *effective* magnetic moment of the electron results from the contributions of the "bare" electron, which is the Dirac particle, and the cloud of "virtual," short-lived electron–positron pairs and photons that surround this particle as a consequence of QED. The small effects of these quantum mechanical fluctuations can be theoretically computed using Feynman diagrams with loops.

The one-loop contribution to the anomalous magnetic moment of the electron, corresponding to the first order and largest correction in QED, is found by calculating the vertex function shown in the diagram above. The calculation was discovered by Julian Schwinger in 1948. Computed to fourth order, the QED prediction for the electron's anomalous magnetic moment agrees with the experimentally measured value to more than 10 significant figures, making the magnetic moment of the electron one of the most accurately verified predictions in the history of physics.

Compared to the electron, the anomalous magnetic moments of the nucleons are enormous. The g-factor for the proton is 5.6, and the chargeless neutron should have no magnetic moment at all. Note, however, that the anomalous magnetic moments of the nucleons, that is, their magnetic moments with the expected Dirac particle magnetic moments subtracted, are roughly equal but of opposite sign: $\mu_p - 1.00\,\mu_N = +1.79\,\mu_N$, $\mu_n - 0.00\,\mu_N = -1.91\,\mu_N$.

The Yukawa interaction for nucleons was discovered in the mid-1930s, and this nuclear force is mediated by pion mesons. In parallel with the theory for the electron, the hypothesis was that higher-order loops involving nucleons and pions may generate the anomalous magnetic moments of the nucleons. The physical picture was that the *effective* magnetic moment of the neutron arose from the combined contributions of the "bare" neutron, which is zero, and the cloud of "virtual" pions and photons that surround this particle as a consequence of the nuclear and electromagnetic forces. The Feynman diagram is roughly the first order diagram, with the role of the virtual particles played by pions. As noted by Abraham Pais, "between late 1948 and the middle of 1949 at least six papers appeared reporting on second order calculations of nucleon moments." These theories were also, as noted by Pais, "a flop" – they gave results that grossly disagreed with observation. Nevertheless, serious efforts continued along these lines for the next couple of decades, to little success. These theoretical approaches were incorrect because the nucleons are composite particles with their magnetic moments arising from their elementary components, quarks.

## Quark Model for Nucleon Magnetic Moments

In the quark model for hadrons, the neutron is composed of one up quark (charge +2/3 *e*) and two down quarks (charge −1/3 *e*). The magnetic moment of the neutron can be modeled as a sum of the magnetic moments of the constituent quarks, although this simple model belies the complexities of the Standard Model of particle physics. The calculation assumes that the quarks behave like pointlike Dirac particles, each having their own magnetic moment, as computed using an expression similar to the one above for the nuclear magneton:

$$\mu_q = \frac{e_q \hbar}{2m_q},$$

where the q-subscripted variables refer to quark magnetic moment, charge, or mass. Simplistically, the magnetic moment of the neutron can be viewed as resulting from the vector sum of the three quark magnetic moments, plus the orbital magnetic moments caused by the movement of the three charged quarks within the neutron.

In one of the early successes of the Standard Model (SU(6) theory), in 1964 Mirza A. B. Beg, Benjamin W. Lee, and Abraham Pais theoretically calculated the ratio of proton to neutron magnetic moments to be −3/2, which agrees with the experimental value to within 3%. The measured value for this ratio is −1.45989806(34). A contradiction of the quantum mechanical basis of this calculation with the Pauli exclusion principle, led to the discovery of the color charge for quarks by Oscar W. Greenberg in 1964.

From the nonrelativistic, quantum mechanical wavefunction for baryons composed of three quarks, a straightforward calculation gives fairly accurate estimates for the magnetic moments of neutrons, protons, and other baryons. For a neutron, the end result of this calculation is that the magnetic moment of the neutron is given by $\mu_n = 4/3\,\mu_d - 1/3\,\mu_u$, where $\mu_d$ and $\mu_u$ are the magnetic moments for the down and up quarks, respectively. This result combines the intrinsic magnetic moments of the quarks with their orbital magnetic moments, and assumes the three quarks are in a particular, dominant quantum state.

| Baryon | Magnetic moment of quark model | Computed ($\mu_N$) | Observed ($\mu_N$) |
|---|---|---|---|
| p | $4/3\,\mu_u - 1/3\,\mu_d$ | 2.79 | 2.793 |
| n | $4/3\,\mu_d - 1/3\,\mu_u$ | −1.86 | −1.913 |

The results of this calculation are encouraging, but the masses of the up or down quarks were assumed to be 1/3 the mass of a nucleon. The masses of the quarks are actually only about 1% that of a nucleon. The discrepancy stems from the complexity of the Standard Model for nucleons, where most of their mass originates in the gluon fields, virtual particles, and their associated energy that are essential aspects of the strong force. Furthermore, the complex system of quarks and gluons that constitute a neutron requires a relativistic treatment. Nucleon magnetic moments have been successfully computed from first principles, requiring significant computing resources.

## Neutron Electric Dipole Moment

The neutron electric dipole moment (nEDM) is a measure for the distribution of positive and negative charge inside the neutron. A finite electric dipole moment can only exist if the centers of the negative and positive charge distribution inside the particle do not coincide. So far, no neutron EDM has been found. The current best upper limit amounts to $|d_n| < 3.0\times10^{-26}\ e\cdot\text{cm}$.

## Theory

A permanent electric dipole moment of a fundamental particle violates both parity (P) and time reversal symmetry (T). This can be understood by examining the neutron with its magnetic dipole moment and hypothetical electric dipole moment. Under time reversal, the magnetic dipole moment changes its direction, whereas the electric dipole moment stays unchanged. Under parity, the electric dipole moment changes its direction but not the magnetic dipole moment. As the resulting system under P and T is not symmetric with respect to the initial system, these symmetries are violated in the case of the existence of an EDM. Having also CPT symmetry, the combined symmetry CP is violated as well.

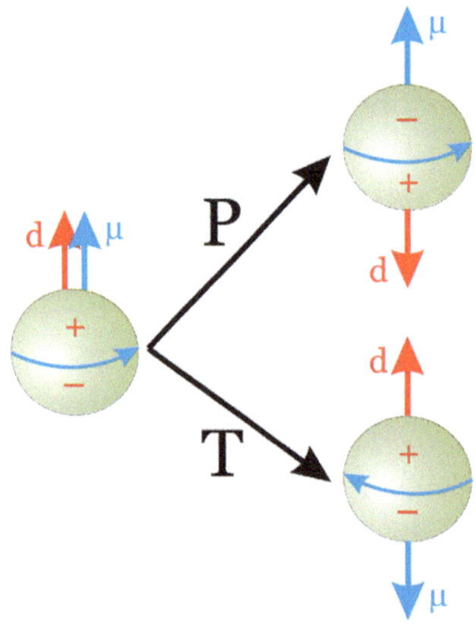


Parity (P) and time reversal (T) violation due to an electric dipole moment

## Standard Model Prediction

As it is depicted above, in order to generate a finite nEDM one needs processes that violate CP symmetry. CP violation has been observed in weak interactions and is included in the Standard Model of particle physics via the CP-violating phase in the CKM matrix. However, the amount of CP violation is very small and therefore also the contribution to the nEDM: $|d_n| \sim 10^{-31}\ e{\cdot}\mathrm{cm}$.

## Matter–antimatter Asymmetry

From the asymmetry between matter and antimatter in the universe, one suspects that there must be a sizeable amount of CP-violation. Measuring a neutron electric dipole moment at a much higher level than predicted by the Standard Model would therefore directly confirm this suspicion and improve our understanding of CP-violating processes.

## Strong CP Problem

As the neutron is built up of quarks, it is also susceptible to CP violation stemming from strong interactions. Quantum chromodynamics – the theoretical description of the strong force – naturally includes a term which breaks CP-symmetry. The strength of this term is characterized by the angle $\theta$. The current limit on the nEDM constrains this angle to be less than $10^{-10}$ rad. This fine-tuning of the $\theta$-angle, which is naturally expected to be of order 1, is the strong CP problem.

## SUSY CP Problem

Supersymmetric extensions to the Standard Model, such as the Minimal Supersymmetric Standard Model, generally lead to a large CP-violation. Typical predictions for the neutron EDM arising from the theory range between $10^{-25}\ e{\cdot}\mathrm{cm}$ and $10^{-28}\ e{\cdot}\mathrm{cm}$. As in the case of the strong interaction, the limit on the neutron EDM is already constraining the CP violating phases. The fine-tuning is, however, not as severe yet.

## Experimental Technique

In order to extract the neutron EDM, one measures the Larmor precession of the neutron spin in the presence of parallel and antiparallel magnetic and electric fields. The precession frequency for each of the two cases is given by

$$h\nu = 2\mu_n B \pm 2d_n E,$$

the addition or subtraction of the frequencies stemming from the precession of the magnetic moment around the magnetic field and the precession of the electric dipole moment around the electric field. From the difference of those two frequencies one readily obtains a measure of the neutron EDM:

$$d_n = \frac{h\Delta\nu}{4E}$$

The biggest challenge of the experiment (and at the same time the source of the biggest systematic false effects) is to ensure that the magnetic field does not change during these two measurements.

## History

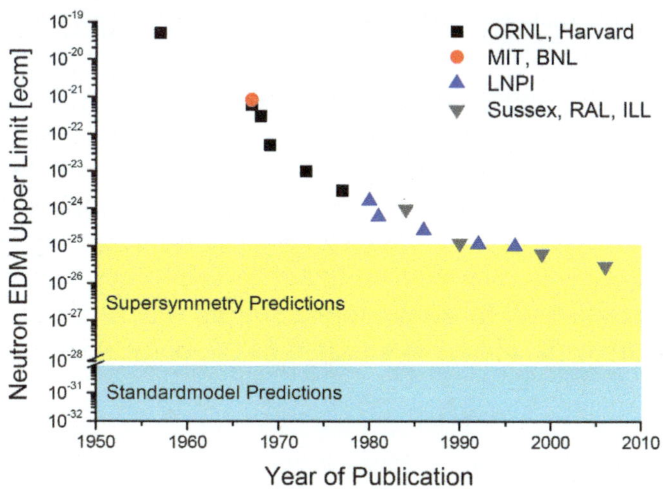


Measured upper limits of the neutron EDM. Given are also the predictions stemming from Supersymmetry and the Standard Model

The first experiments searching for the electric dipole moment of the neutron used beams of thermal (and later cold) neutrons to conduct the measurement. It started with the experiment by Smith, Purcell and Ramsey in 1951 (and published in 1957) obtaining a limit of $|d_n| < 5\times10^{-20}$ $e$·cm. Beams of neutrons were used until 1977 for nEDM experiments. At this point, systematic effects related to the high velocities of the neutrons in the beam became insurmountable. The final limit obtained with a neutron beam amounts to $|d_n| < 3\times10^{-24}$ $e$·cm.

After that, experiments with ultracold neutrons took over. It started in 1980 with an experiment at the Leningrad Nuclear Physics Institute obtaining a limit of $|d_n| < 1.6\times10^{-24}$ $e$·cm. This experiment and especially the experiment starting in 1984 at the Institut Laue-Langevin pushed the limit down by another two orders of magnitude yielding the above quoted best upper limit in 2006, revised in 2015.

During these 50 years of experiments, six orders of magnitude have been covered thereby putting stringent constraints on theoretical models.

## Current Experiments

Currently, there are at least five experiments aiming at improving the current limit (or measuring for the first time) on the neutron EDM with a sensitivity down to $10^{-28}$ $e{\cdot}$cm over the next 10 years, thereby covering the range of prediction coming from supersymmetric extensions to the Standard Model.

- nEDM experiment running (n2EDM under construction) at the UCN source at the Paul Scherrer Institute
- UCN nEDM experiment under construction at TRIUMF
- Cryogenic neutron EDM experiment being set up at the Institut Laue-Langevin
- nEDM experiment being envisaged at the Spallation Neutron Source
- nEDM experiment being built at the Institut Laue-Langevin
- nEDM experiment being built at the Forschungsreaktor München II

## Shape of the Nucleus: Electric Moments and magnetic Moment

The shape of a nucleus is an important field of study. Some nuclei are almost spherical, whereas some are deformed. Determination of degree of deformation of nucleus both theoretically and experimentally is a complicated topic. An initial idea of deformation can be obtained from electric quadrupole moment of a nucleus. The electric quadrupole moment measures the departure of a nucleus from spherical symmetry. Here we adopt a simple way to determine electric quadrupole moment. Let us consider that the charge is situated at the point S having Cartesian co-ordinates (x,y,z).

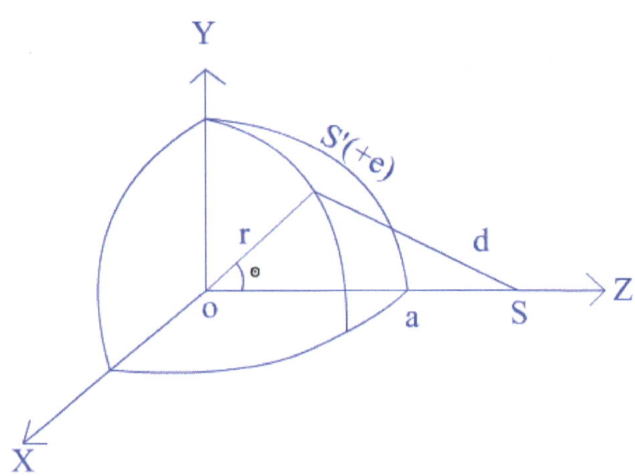


Coordinate system for evaluation of potential

The potential at a point S on the Z-axis due to this charge is:

$$\phi = \frac{e}{d}$$

Where $d = (a^2 + r^2 - 2ar\cos\theta)^{\frac{1}{2}}$

where r is the distance of the charge from the origin and is given by $r = (x^2 + y^2 + z^2)^{1/2}$ and $\cos\theta = z/r$ defines the angle between a and r.

$$\phi = \frac{e}{(a^2 + r^2 - 2ar\cos\theta)^{\frac{1}{2}}} = \frac{e\left(1 - \frac{2r\cos\theta}{a} + \frac{r^2}{a^2}\right)^{\frac{-1}{2}}}{a}$$

$$\theta = \frac{e}{a} + \frac{er\cos\theta}{a^2} + \frac{er^2\left(\frac{3\cos^2\theta}{2} - \frac{1}{2}\right)}{a^3} + h.o.t$$

In the equation above, the coefficient of $1/a$ is known as monopole strength, the coefficient of $1/a^2$ is the z-component of the dipole moment, the coefficient of $1/a^3$ is the z-component of the quadrupole moment. Similarly, considering higher order terms (h.o.t.) one get higher moments like octupole moments, etc. The first term in equation is the ordinary Coulomb potential.

The net quadrupole moment of the nucleus is:

$$Q_d = er^2\left(\frac{3\cos^2\theta}{2} - \frac{1}{2}\right)$$

Putting $\cos\theta = z/r$, we get

$$Q_d = \frac{e}{2}(3z^2 - r^2)$$

From electrostatics, it is well known that the electric field due to a uniformly charged shell at all points outside of it is the same as if all the charge were concentrated at the centre of the shell, i.e the centre of the nucleus here. Therefore, from equation one can easily find out that a nucleus with a spherical symmetric charge distribution has no electric quadrupole moment or higher electric moments. If a uniformly charged ellipsoid is considered than it has both a net charge and a quadrupole moment, but its dipole moment is zero. Its charge distribution can be well approximated, as shown in the figure, by a spherical charge distribution plus an elementary quadrupole.

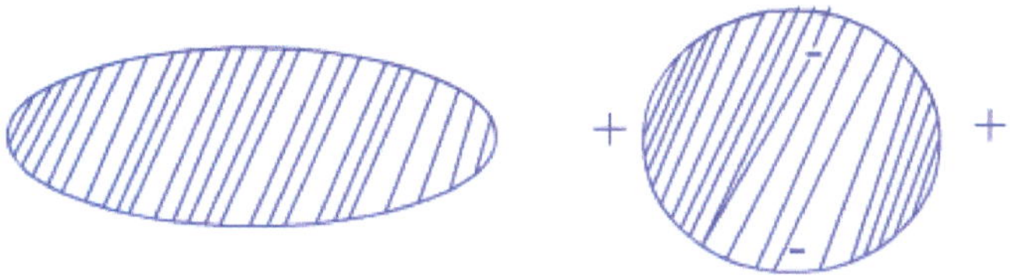


A uniformly positively charged distribution of ellipsoidal shape (at the left), which is electrically equivalent to a uniform positively charged sphere plus extra positive and negative charges as shown at the right.

From a detail quantum mechanical analysis, one can show that for a spherical distribution of charges Q = 0, and Q > 0 and Q < 0 are for prolate (cigar-like shape) and oblate distribution (discus-shape) respectively.

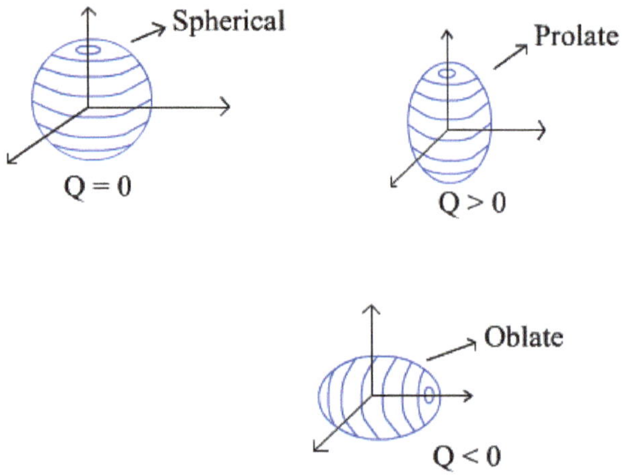


Pictorial representation of spherical, prolate and oblate shape nucleus

## Nucleus - A Tiny Magnet

We have seen that nucleus has a charge distribution and various electrical moments. Apart from charge, nucleus also has a spin,* which essentially makes nucleus to behave like a tiny magnet with well defined magnetic moment. Every nucleon inside a nucleus has a magnetic moment due to its spin motion. The magnetic moment can be calculated considering a classical orbital motion of a particle, one can write the angular momentum

$$\check{l} = \vec{r} \times \vec{p}$$

where $\vec{r}$ and $\vec{p}$ are radius of the orbit and linear momentum of the particle respectively. If the current produced by the motion of the charge is equal to i then the magnetic moment (μ) associated with it is

$$\vec{\mu} = \pi r^2 l \acute{i}$$

Where $\acute{i}$ is a unit vector vertical to the circular motion, in the rotation sense going with a positive current. For a proton or electron combining the above equations one can write

$$\left|\vec{\mu}\right| = \frac{e}{2m}\left|\vec{l}\right|$$

In vector notation

$$\vec{\mu} = \frac{e}{2m}\acute{l}$$

Quantum mechanically one can write the relation between the operators for the above circular motion

$$\hat{\mu} = \frac{e}{2m}\hat{l}$$

And

$$\hat{\mu}_{l,x} = \frac{e}{2m}\hat{l}_x$$

the z component of the magnetic moment.

From the quantum mechanical consideration,2 Equation can be written as:

$$\hat{\mu}_{l,x} = \frac{e\hbar}{2m}m_l$$

The unit $\frac{e\hbar}{2m}$ is known as Bohr (for electron) magneton or nuclear (if m is the nucleon mass) magneton. Therefore,

$$\mu_{l,x} = m_l \mu_N$$

For intrinsic spin an analogous relationship can be written as

$$\mu_{s,x} = g_s \mu_N m_s$$

where $g_s$ is the gyromagnetic ratio. For electron the gyromagnetic ratio turns out to be almost -2 and for protons and neutrons 5.5855 and -3.8263 respectively. The individual moments can be combined and the total moment can be written as

$$\mu_{J,x} = g_j \mu_N m_J$$

where $g_j$ is the nuclear gyromagnetic ratio. The magnetic moments of electron proton and neutron are

$$\mu_e = \frac{eh}{4\pi m_e}$$

$$\mu_p = \frac{2.7925eh}{4\pi m_p}$$

And

$$\mu_n = \frac{-1.9128eh}{4\pi m_p}$$

respectively. The numerical factors appear in case of nucleons is due to the fact that neutrons and protons have electric charge distribution i.e. of charged meson clouds surrounding them.

# Nuclear Binding Energy

Nuclear binding energy is the energy that would be required to disassemble the nucleus of an atom into its component parts. These component parts are neutrons and protons, which are collectively called nucleons. The binding energy of nuclei is due to the attractive forces that hold these nucleons together, and it is always a positive number, since all nuclei would require the expenditure of energy to separate them into individual protons and neutrons. The mass of an atomic nucleus is less than the sum of the individual masses of the free constituent protons and neutrons (according to Einstein's equation $E=mc^2$) and this 'missing mass' is known as the mass defect, and represents the energy that was released when the nucleus was formed.

The term "nuclear binding energy" may also refer to the energy balance in processes in which the nucleus splits into fragments composed of more than one nucleon. If new binding energy is available when light nuclei fuse, or when heavy nuclei split, either process can result in release of this binding energy. This energy may be made available as *nuclear energy* and can be used to produce electricity as in (nuclear power) or in a nuclear weapon. When a large nucleus splits into pieces, excess energy is emitted as photons (gamma rays) and as the kinetic energy of a number of different ejected particles (nuclear fission products).

The nuclear binding energies and forces are on the order of a million times greater than the electron binding energies of light atoms like hydrogen.

The mass defect of a nucleus represents the mass of the energy of binding of the nucleus, and is the difference between the mass of a nucleus and the sum of the masses of the nucleons of which it is composed.

## Introduction

Nuclear binding energy is explained by the basic principles involved in nuclear physics.

## Nuclear Energy

An absorption or release of nuclear energy occurs in nuclear reactions or radioactive decay; those that absorb energy are called endothermic reactions and those that release energy are exothermic reactions. Energy is consumed or liberated because of differences in the nuclear binding energy between the incoming and outgoing products of the nuclear transmutation.

The best-known classes of exothermic nuclear transmutations are fission and fusion. Nuclear energy may be liberated by atomic fission, when heavy atomic nuclei (like uranium and plutonium) are broken apart into lighter nuclei. The energy from fission is used to generate electric power in hundreds of locations worldwide. Nuclear energy is also released during atomic fusion, when light nuclei like hydrogen are combined to form heavier nuclei such as helium. The Sun and other stars use nuclear fusion to generate thermal energy which is later radiated from the surface, a type of stellar nucleosynthesis. In any exothermic nuclear process, nuclear mass might ultimately be converted to thermal energy, given off as heat.

In order to quantify the energy released or absorbed in any nuclear transmutation, one must know the nuclear binding energies of the nuclear components involved in the transmutation.

## The Nuclear Force

Electrons and nuclei are kept together by electrostatic attraction (negative attracts positive). Furthermore, electrons are sometimes shared by neighboring atoms or transferred to them (by processes of quantum physics), and this link between atoms is referred to as a chemical bond, and is responsible for the formation of all chemical compounds.

The force of electric attraction does not hold nuclei together, because all protons carry a positive charge and repel each other. Thus, electric forces do not hold nuclei together, because they act in the opposite direction. It has been established that binding neutrons to nuclei clearly requires a non-electrical attraction.

Therefore, another force, called the *nuclear force* (or *residual strong force*) holds the nucleons of nuclei together. This force is a residuum of the strong interaction, which binds quarks into nucleons at an even smaller level of distance.

The nuclear force must be stronger than the electric repulsion at short distances, but weaker far away, or else different nuclei might tend to clump together. Therefore, it has short-range characteristics. An analogy to the nuclear force is the force between two small magnets: magnets are very difficult to separate when stuck together, but once pulled a short distance apart, the force between them drops almost to zero.

Unlike gravity or electrical forces, the nuclear force is effective only at very short distances. At greater distances, the electrostatic force dominates: the protons repel each other because they are positively charged, and like charges repel. For that reason, the protons forming the nuclei of ordinary hydrogen—for instance, in a balloon filled with hydrogen—do not combine to form helium (a process that also would require some protons to combine with electrons and become neutrons). They cannot get close enough for the nuclear force, which attracts them to each other, to become important. Only under conditions of extreme pressure and temperature (for example, within the core of a star), can such a process take place.

## Physics of Nuclei

The nuclei of atoms are found in many different sizes. In hydrogen they contain just one proton, in deuterium or heavy hydrogen a proton and a neutron; in helium, two protons and two neutrons, and in carbon, nitrogen and oxygen - six, seven and eight of each particle, respectively. A helium nucleus weighs less than the sum of the weights of its components. The same phenomenon is found for carbon, nitrogen and oxygen. For example, the carbon nucleus is slightly lighter than three helium nuclei, which can combine to make a carbon nucleus. This illustrates the mass defect.

## Mass Defect

Mass defect is the difference between the mass of a composite particle and the sum of the masses of its parts.

The “mass defect” can be explained using Albert Einstein’s formula $E = m c^2$, expressing the equivalence of energy and mass. By this formula, adding energy also increases mass (both weight and inertia), whereas removing energy decreases mass.

If a combination of particles contains extra energy—for instance, in a molecule of the explosive TNT—weighing it reveals some extra mass, compared to its end products after an explosion. (The weighing must be done after the products have been stopped and cooled, however, as the extra mass must escape from the system as heat before its loss can be noticed, in theory.) On the other hand, if one must inject energy to separate a system of particles into its components, then the initial mass is less than that of the components after they are separated. In the latter case, the energy injected is "stored" as potential energy, which shows as the increased mass of the components that store it. This is an example of the fact that energy of all types is seen in systems as mass, since mass and energy are equivalent, and each is a "property" of the other.

The latter scenario is the case with nuclei such as helium: to break them up into protons and neutrons, one must inject energy. On the other hand, if a process existed going in the opposite direction, by which hydrogen atoms could be combined to form helium, then energy would be released. The energy can be computed using $E = \Delta m c^2$ for each nucleus, where $\Delta m$ is the difference between the mass of the helium nucleus and the mass of four protons (plus two electrons, absorbed to create the neutrons of helium).

For elements heavier than oxygen, the energy that can be released by assembling them from lighter elements decreases, up to iron. For nuclei heavier than iron, one actually releases energy by breaking them up into 2 fragments. That is how energy is extracted by breaking up uranium nuclei in nuclear power reactors.

The reason the trend reverses after iron is the growing positive charge of the nuclei. The electric force may be weaker than the nuclear force, but its range is greater: in an iron nucleus, each proton repels the other 25 protons, while the nuclear force only binds close neighbors.

As nuclei grow bigger still, this disruptive effect becomes steadily more significant. By the time polonium is reached (84 protons), nuclei can no longer accommodate their large positive charge, but emit their excess protons quite rapidly in the process of alpha radioactivity—the emission of helium nuclei, each containing two protons and two neutrons. (Helium nuclei are an especially stable combination.) Because of this process, nuclei with more than 94 protons are not found naturally on Earth. The isotopes beyond uranium (atomic number 92) with the longest half-lives are plutonium-244 (80 million years) and curium-247 (16 million years).

## Solar Binding Energy

The nuclear fusion process works as follows: five billion years ago, the new Sun formed when gravity pulled together a vast cloud of hydrogen and dust, from which the Earth and other planets also arose. The gravitational pull released energy and heated the early Sun, much in the way Helmholtz proposed.

Thermal energy appears as the motion of atoms and molecules: the higher the temperature of a collection of particles, the greater is their velocity and the more violent are their collisions. When the temperature at the center of the newly formed Sun became great enough for collisions between hydrogen nuclei to overcome their electric repulsion, and bring them into the short range of the attractive nuclear force, nuclei began to stick together. When this began to happen, protons combined into deuterium and then helium, with some protons changing in the process to neutrons (plus positrons, positive electrons, which combine with electrons and are both destroyed). This released nuclear energy now keeps up the high temperature of the Sun's core, and the heat also keeps the gas pressure high, keeping the Sun at its present size, and stopping gravity from compressing it any more. There is now a stable balance between gravity and pressure.

Different nuclear reactions may predominate at different stages of the Sun's existence, including the proton-proton reaction and the carbon-nitrogen cycle—which involves heavier nuclei, but whose final product is still the combination of protons to form helium.

A branch of physics, the study of controlled nuclear fusion, has tried since the 1950s to derive useful power from nuclear fusion reactions that combine small nuclei into bigger ones, typically to heat boilers, whose steam could turn turbines and produce electricity. Unfortunately, no earthly laboratory can match one feature of the solar powerhouse: the great mass of the Sun, whose weight keeps the hot plasma compressed and confines the nuclear furnace to the Sun's core. Instead, physicists use strong magnetic fields to confine the plasma, and for fuel they use heavy forms of hydrogen, which burn more easily. Magnetic traps can be rather unstable, and any plasma hot enough and dense enough to undergo nuclear fusion tends to slip out of them after a short time. Even with ingenious tricks, the confinement in most cases lasts only a small fraction of a second.

## Combining Nuclei

Small nuclei that are larger than hydrogen can combine into bigger ones and release energy, but in combining such nuclei, the amount of energy released is much smaller compared to hydrogen fusion. The reason is that while the overall process releases energy from letting the nuclear attraction do its work, energy must first be injected to force together positively charged protons, which also repel each other with their electric charge.

For elements that weigh more than iron (a nucleus with 26 protons), the fusion process no longer releases energy. In even heavier nuclei energy is consumed, not released, by combining similar sized nuclei. With such large nuclei, overcoming the electric repulsion (which affects all protons in the nucleus) requires more energy than what is released by the nuclear attraction (which is effective mainly between close neighbors). Conversely, energy could actually be released by breaking apart nuclei heavier than iron.

With the nuclei of elements heavier than lead, the electric repulsion is so strong that some of them spontaneously eject positive fragments, usually nuclei of helium that form very stable combinations (alpha particles). This spontaneous break-up is one of the forms of radioactivity exhibited by some nuclei.

Nuclei heavier than lead (except for bismuth, thorium, uranium, and plutonium) spontaneously break up too quickly to appear in nature as primordial elements, though they can be produced artificially or as intermediates in the decay chains of lighter elements. Generally, the heavier the nuclei are, the faster they spontaneously decay.

Iron nuclei are the most stable nuclei (in particular iron-56), and the best sources of energy are therefore nuclei whose weights are as far removed from iron as possible. One can combine the lightest ones—nuclei of hydrogen (protons)—to form nuclei of helium, and that is how the Sun generates its energy. Or else one can break up the heaviest ones—nuclei of uranium or plutonium—into smaller fragments, and that is what nuclear power reactors do.

## Nuclear Binding Energy

An example that illustrates nuclear binding energy is the nucleus of $^{12}C$ (carbon-12), which contains 6 protons and 6 neutrons. The protons are all positively charged and repel each other, but the nuclear force overcomes the repulsion and causes them to stick together. The nuclear force is a close-range force (it is strongly attractive at a distance of 1.0 fm and becomes extremely small beyond a distance of 2.5fm), and virtually no effect of this force is observed outside the nucleus. The nuclear force also pulls neutrons together, or neutrons and protons.

The energy of the nucleus is negative with regard to the energy of the particles pulled apart to infinite distance (just like the gravitational energy of planets of the solar system), because energy must be utilized to split a nucleus into its individual protons and neutrons. Mass spectrometers have measured the masses of nuclei, which are always less than the sum of the masses of protons and neutrons that form them, and the difference—by the formula $E = m\,c^2$—gives the binding energy of the nucleus.

## Nuclear Fusion

The binding energy of helium is the energy source of the Sun and of most stars. The sun is composed of 74 percent hydrogen (measured by mass), an element whose nucleus is a single proton. Energy is released in the sun when 4 protons combine into a helium nucleus, a process in which two of them are also converted to neutrons.

The conversion of protons to neutrons is the result of another nuclear force, known as the weak (nuclear) force. The weak force, like the strong force, has a short range, but is much weaker than the strong force. The weak force tries to make the number of neutrons and protons into the most energetically stable configuration. For nuclei containing less than 40 particles, these numbers are usually about equal. Protons and neutrons are closely related and are sometimes collectively known as nucleons. As the number of particles increases toward a maximum of about 209, the number of neutrons to maintain stability begins to outstrip the number of protons, until the ratio of neutrons to protons is about three to two.

The protons of hydrogen combine to helium only if they have enough velocity to overcome each other's mutual repulsion sufficiently to get within range of the strong nuclear attraction. This means that fusion only occurs within a very hot gas. Hydrogen hot enough for combining to helium requires an enormous pressure to keep it confined, but suitable conditions exist in the central regions of the Sun, where such pressure is provided by the enormous weight of the layers above the core, pressed inwards by the Sun's strong gravity. The process of combining protons to form helium is an example of nuclear fusion.

The earth's oceans contain a large amount of hydrogen that could theoretically be used for fusion, and helium byproduct of fusion does not harm the environment, so some consider nuclear fusion a good alternative to supply humanity's energy needs. Experiments to generate electricity from fusion have so far have only partially succeeded. Sufficiently hot hydrogen must be ionized and confined. One technique is to use very strong magnetic fields, because charged particles (like those trapped in the Earth's radiation belt) are guided by magnetic field lines. Fusion experiments also rely on heavy hydrogen, which fuses more easily, and gas densities can be moderate. But even with these techniques far more net energy is consumed by the fusion experiments than is yielded by the process.

## The Binding Energy Maximum and Ways to Approach it by Decay

In the main isotopes of light nuclei, such as carbon, nitrogen and oxygen, the most stable combination of neutrons and of protons are when the numbers are equal (this continues to element 20, calcium). However, in heavier nuclei, the disruptive energy of protons increases, since they are confined to a tiny volume and repel each other. The energy of the strong force holding the nucleus together also increases, but at a slower rate, as if inside the nucleus, only nucleons close to each other are tightly bound, not ones more widely separated.

The net binding energy of a nucleus is that of the nuclear attraction, minus the disruptive energy of the electric force. As nuclei get heavier than helium, their net binding energy per nucleon (deduced from the difference in mass between the nucleus and the sum of masses of component nucleons) grows more and more slowly, reaching its peak at iron. As nucleons are added, the total nuclear binding energy always increases—but the total disruptive energy of electric forces (positive protons repelling other protons) also increases, and past iron, the second increase outweighs the first. Iron-56 ($^{56}Fe$) is the most efficiently bound nucleus meaning that it has the least average mass per nucleon. However, nickel-62 is the most tightly bound nucleus in terms of energy of binding per nucleon. (Nickel-62's higher energy of binding does not translate to a larger mean mass loss than Fe-56, because Ni-62 has a slightly higher ratio of neutrons/protons than does iron-56, and the presence of the heavier neutrons increases nickel-62's average mass per nucleon).

To reduce the disruptive energy, the weak interaction allows the number of neutrons to exceed that of protons—for instance, the main isotope of iron has 26 protons and 30 neutrons. Isotopes also exist where the number of neutrons differs from the most stable number for that number of nucleons. If the ratio of protons to neutrons is too far from stability, nucleons may spontaneously change from proton to neutron, or neutron to proton.

The two methods for this conversion are mediated by the weak force, and involve types of beta decay. In the simplest beta decay, neutrons are converted to protons by emitting a negative electron and an antineutrino. This is always possible outside a nucleus because neutrons are more massive than protons by an equivalent of about 2.5 electrons. In the opposite process, which only happens within a nucleus, and not to free particles, a proton may become a neutron by ejecting a positron. This is permitted if enough energy is available between parent and daughter nuclides to do this (the required energy difference is equal to 1.022 MeV, which is the mass of 2 electrons). If the mass difference between parent and daughter is less than this, a proton-rich nucleus may still convert protons to neutrons by the process of electron capture, in which a proton simply electron captures one of the atom's K orbital electrons, emits a neutrino, and becomes a neutron.

Among the heaviest nuclei, starting with tellurium nuclei (element 52) containing 106 or more nucleons, electric forces may be so destabilizing that entire chunks of the nucleus may be ejected, usually as alpha particles, which consist of two protons and two neutrons (alpha particles are fast helium nuclei). (Beryllium-8 also decays, very quickly, into two alpha particles.) Alpha particles are extremely stable. This type of decay becomes more and more probable as elements rise in atomic weight past 106.

The curve of binding energy is a graph that plots the binding energy per nucleon against atomic mass. This curve has its main peak at iron and nickel and then slowly decreases again, and also a narrow isolated peak at helium, which as noted is very stable. The heaviest nuclei in nature, uranium $^{238}U$, are unstable, but having a half-life of 4.5 billion years, close to the age of the Earth, they are still relatively abundant; they (and other nuclei heavier than iron) may have formed in a supernova explosion preceding the formation of the solar system. The most common isotope of thorium, $^{232}Th$, also undergoes alpha particle emission, and its half-life (time over which half a number of atoms decays) is even longer, by several times. In each of these, radioactive decay produces daughter isotopes that are also unstable, starting a chain of decays that ends in some stable isotope of lead.

## Determining Nuclear Binding Energy

Calculation can be employed to determine the nuclear binding energy of nuclei. The calculation involves determining the *mass defect*, converting it into energy, and expressing the result as energy per mole of atoms, or as energy per nucleon.

## Conversion of Mass Defect into Energy

Mass defect is defined as the difference between the mass of a nucleus, and the sum of the masses of the nucleons of which it is composed. The mass defect is determined by calculating three quantities. These are: the actual mass of the nucleus, the composition of the nucleus (number of protons and of neutrons), and the masses of a proton and of a neutron. This is then followed by converting the mass defect into energy. This quantity is the nuclear binding energy, however it must be expressed as energy per mole of atoms or as energy per nucleon.

## Fission and Fusion

Nuclear energy is released by the splitting (fission) or merging (fusion) of the nuclei of atom(s). The conversion of nuclear mass-energy to a form of energy, which can remove some mass when the energy is removed, is consistent with the mass-energy equivalence formula:

$\Delta E = \Delta m\, c^2$,

in which,

$\Delta E$ = energy release,

$\Delta m$ = mass defect,

and $c$ = the speed of light in a vacuum (a physical constant).

Nuclear energy was first discovered by French physicist Henri Becquerel in 1896, when he found that photographic plates stored in the dark near uranium were blackened like X-ray plates (X-rays had recently been discovered in 1895).

Nuclear chemistry can be used as a form of alchemy to turn lead into gold or change any atom to any other atom (though this may require many intermediate steps). Radionuclide (radioisotope) production often involves irradiation of another isotope (or more precisely a nuclide), with alpha particles, beta particles, or gamma rays. Nickel-62 has the highest binding energy per nucleon of any isotope. If an atom of lower average binding energy is changed into two atoms of higher average binding energy, energy is given off. Also, if two atoms of lower average binding energy fuse into an atom of higher average binding energy, energy is given off. The chart shows that fusion of hydrogen, the combination to form heavier atoms, releases energy, as does fission of uranium, the breaking up of a larger nucleus into smaller parts. Stability varies between isotopes: the isotope U-235 is much less stable than the more common U-238.

Nuclear energy is released by three *exoenergetic* (or exothermic) processes:

- Radioactive decay, where a neutron or proton in the radioactive nucleus decays spontaneously by emitting either particles, electromagnetic radiation (gamma rays), or both. Note that for radioactive decay, it is not strictly necessary for the binding energy to increase. What is strictly necessary is that the mass decrease. If a neutron turns into a proton and the energy of the decay is less than 0.782343 MeV (such as rubidium-87 decaying to strontium-87), the average binding energy per nucleon will actually decrease.
- Fusion, two atomic nuclei fuse together to form a heavier nucleus.
- Fission, the breaking of a heavy nucleus into two (or more rarely three) lighter nuclei.

## Binding Energy for Atoms

The binding energy of an atom (including its electrons) is not the same as the binding energy of the atom's nucleus. The measured mass deficits of isotopes are always listed as mass deficits of the neutral atoms of that isotope, and mostly in MeV. As a consequence, the listed mass deficits are not a measure for the stability or binding energy of isolated nuclei, but for the whole atoms. This has very practical reasons, because it is very hard to totally ionize heavy elements, i.e. strip them of all of their electrons.

This practice is useful for other reasons, too: stripping all the electrons from a heavy unstable nucleus (thus producing a bare nucleus) changes the lifetime of the nucleus, indicating that the nucleus cannot be treated independently (Experiments at the heavy ion accelerator GSI). This is also evident from phenomena like electron capture. Theoretically, in orbital models of heavy atoms, the electron orbits partially inside the nucleus (it doesn't *orbit* in a strict sense, but has a non-vanishing probability of being located inside the nucleus).

A nuclear decay happens to the nucleus, meaning that properties ascribed to the nucleus change in the event. In the field of physics the concept of "mass deficit" as a measure for "binding energy" means "mass deficit of the neutral atom" (not just the nucleus) and is a measure for stability of the whole atom.

## Nuclear Binding Energy Curve

The region of increasing binding energy is followed by a region of relative stability (saturation) in the sequence from magnesium through xenon. In this region, the nucleus has become large enough that nuclear forces no longer completely extend efficiently across its width. Attractive nuclear forces in this region, as atomic mass increases, are nearly balanced by repellent electromagnetic forces between protons, as the atomic number increases.

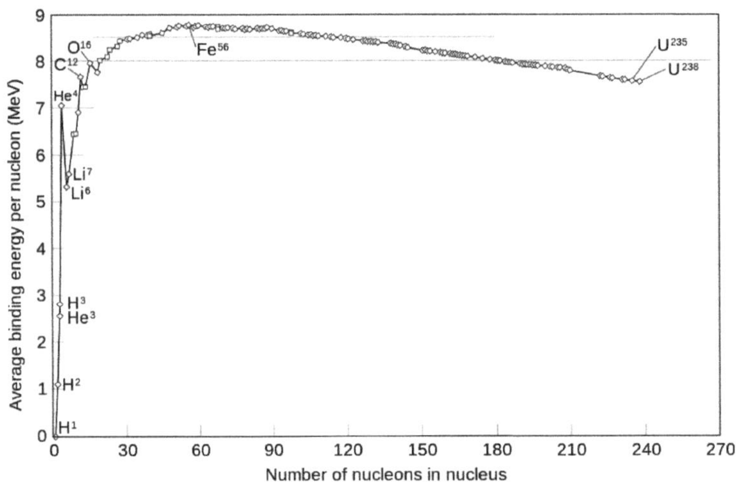


In the periodic table of elements, the series of light elements from hydrogen up to sodium is observed to exhibit generally increasing binding energy per nucleon as the atomic mass increases. This increase is generated by increasing forces per nucleon in the nucleus, as each additional nucleon is attracted by other nearby nucleons, and thus more tightly bound to the whole.

Finally, in elements heavier than xenon, there is a decrease in binding energy per nucleon as atomic number increases. In this region of nuclear size, electromagnetic repulsive forces are beginning to overcome the strong nuclear force attraction.

At the peak of binding energy, nickel-62 is the most tightly bound nucleus (per nucleon), followed by iron-58 and iron-56. This is the approximate basic reason why iron and nickel are very common metals in planetary cores, since they are produced profusely as end products in supernovae and in the final stages of silicon burning in stars. However, it is not binding energy per defined nucleon (as defined above), which controls which exact nuclei are made, because within stars, neutrons are free to convert to protons to release even more energy, per generic nucleon, if the result is a stable nucleus with a larger fraction of protons. In fact, it has been argued that photodisintegration of $^{62}$Ni to form $^{56}$Fe may be energetically possible in an extremely hot star core, due to this beta decay conversion of neutrons to protons. The conclusion is that at the pressure and temperature conditions in the cores of large stars, energy is released by converting all matter into $^{56}$Fe nuclei (ionized atoms). (However, at high temperatures not all matter will be in the lowest energy state.) This energetic maximum should also hold for ambient conditions, say $T = 298$ K and $p = 1$ atm, for neutral condensed matter consisting of $^{56}$Fe atoms—however, in these conditions nuclei of atoms are inhibited from fusing into the most stable and low energy state of matter.

It is generally believed that iron-56 is more common than nickel isotopes in the universe for mechanistic reasons, because its unstable progenitor nickel-56 is copiously made by staged build-up of 14 helium nuclei inside supernovas, where it has no time to decay to iron before being released into the interstellar medium in a matter of a few minutes, as the supernova explodes. However, nickel-56 then decays to cobalt-56 within a few weeks, then this radioisotope finally decays to iron-56 with a half life of about 77.3 days. The radioactive decay-powered light curve of such a process has been observed to happen in type II supernovae, such as SN 1987A. In a star, there are no good ways to create nickel-62 by alpha-addition processes, or else there would presumably be more of this highly stable nuclide in the universe.

## Binding Energy and Nuclide Masses

The fact that the maximum binding energy is found in medium-sized nuclei is a consequence of the trade-off in the effects of two opposing forces that have different range characteristics. The attractive nuclear force (strong nuclear force), which binds protons and neutrons equally to each other, has a limited range due to a rapid exponential decrease in this force with distance. However, the repelling electromagnetic force, which acts between protons to force nuclei apart, falls off with distance much more slowly (as the inverse square of distance). For nuclei larger than about four nucleons in diameter, the additional repelling force of additional protons more than offsets any binding energy that results between further added nucleons as a result of additional strong force interactions. Such nuclei become increasingly less tightly bound as their size increases, though most of them are still stable. Finally, nuclei containing more than 209 nucleons (larger than about 6 nucleons in diameter) are all too large to be stable, and are subject to spontaneous decay to smaller nuclei.

Nuclear fusion produces energy by combining the very lightest elements into more tightly bound elements (such as hydrogen into helium), and nuclear fission produces energy by splitting the heaviest elements (such as uranium and plutonium) into more tightly bound elements (such as barium and krypton). Both processes produce energy, because middle-sized nuclei are the most tightly bound of all.

As seen above in the example of deuterium, nuclear binding energies are large enough that they may be easily measured as fractional mass deficits, according to the equivalence of mass and energy. The atomic binding energy is simply the amount of energy (and mass) released, when a collection of free nucleons are joined together to form a nucleus.

Nuclear binding energy can be computed from the difference in mass of a nucleus, and the sum of the masses of the number of free neutrons and protons that make up the nucleus. Once this mass difference, called the mass defect or mass deficiency, is known, Einstein's mass-energy equivalence formula $E = mc^2$ can be used to compute the binding energy of any nucleus. Early nuclear physicists used to refer to computing this value as a «packing fraction» calculation.

For example, the atomic mass unit (1 u) is defined as 1/12 of the mass of a $^{12}$C atom—but the atomic mass of a $^{1}$H atom (which is a proton plus electron) is 1.007825 $u$, so each nucleon in $^{12}$C has lost, on average, about 0.8% of its mass in the form of binding energy.

## Semiempirical Formula for Nuclear Binding Energy

For a nucleus with $A$ nucleons, including $Z$ protons and $N$ neutrons, a semi-empirical formula for the binding energy (BE) per nucleon is:

$$\frac{\text{BE}}{A \cdot \text{MeV}} = a - \frac{b}{A^{1/3}} - \frac{cZ^2}{A^{4/3}} - \frac{d(N-Z)^2}{A^2} \pm \frac{e}{A^{7/4}}$$

where the coefficients are given by: $a = 14.0$; $b = 13.0$; $c = 0.585$; $d = 19.3$; $e = 33$.

The first term $a$ is called the saturation contribution and ensures that the binding energy per nucleon is the same for all nuclei to a first approximation. The term $-b / A^{1/3}$ is a surface tension effect and is proportional to the number of nucleons that are situated on the nuclear surface; it is largest for light nuclei. The term $-cZ^2 / A^{4/3}$ is the Coulomb electrostatic repulsion; this becomes more important as $Z$ increases. The symmetry correction term $-d(N-Z)^2 / A^2$ takes into account the fact that in the absence of other effects the most stable arrangement has equal numbers of protons and neutrons; this is because the n-p interaction in a nucleus is stronger than either the n-n or p-p interaction. The pairing term $\pm e / A^{7/4}$ is purely empirical; it is + for even-even nuclei and - for odd-odd nuclei.

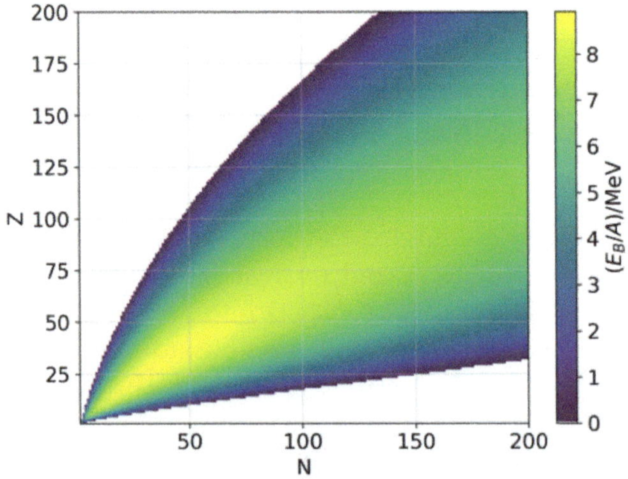


A graphical representation of the semi-empirical binding energy formula. The binding energy per nucleon in MeV (highest numbers in yellow, in excess of 8.5 MeV per nucleon) is plotted for various nuclides as a function of $Z$, the atomic number (y-axis), vs. $N$, the number of neutrons (x-axis). The highest numbers are seen for $Z = 26$ (iron).

## Example Values Deduced from Experimentally Measured Atom Nuclide Masses

The following table lists some binding energies and mass defect values. Notice also that we use 1 u = (931.494028 ± 0.000023) MeV. To calculate the binding energy we use the formula $Z(m_p + m_e) + N m_n - m_{nuclide}$ where $Z$ denotes the number of protons in the nuclides and $N$ their number of neutrons. We take $m_p = 938.2723$ MeV, $m_e = 0.5110$ MeV and $m_n = 939.5656$ MeV. The letter $A$ denotes the sum of $Z$ and $N$ (number of nucleons in the nuclide). If we assume the reference nucleon has the mass of a neutron (so that all "total" binding energies calculated are maximal) we could define the total binding energy as the difference from the mass of the nucleus, and the mass of a collection of $A$ free neutrons. In other words, it would be $(Z + N) m_n - m_{nuclide}$. The "*total* binding energy per nucleon" would be this value divided by $A$.

| Most strongly bound nuclides atoms | | | | | | | | | |
|---|---|---|---|---|---|---|---|---|---|
| **nuclide** | ***Z*** | ***N*** | **mass excess** | **total mass** | **total mass / *A*** | **total binding energy / *A*** | **mass defect** | **binding energy** | **binding energy / *A*** |
| $^{56}$Fe | 26 | 30 | -60.6054 MeV | 55.934937 u | 0.9988372 u | 9.1538 MeV | 0.528479 u | 492.275 MeV | 8.7906 MeV |
| $^{58}$Fe | 26 | 32 | -62.1534 MeV | 57.932276 u | 0.9988496 u | 9.1432 MeV | 0.547471 u | 509.966 MeV | 8.7925 MeV |

| | | | | | | | | | |
|---|---|---|---|---|---|---|---|---|---|
| $^{60}$Ni | 28 | 32 | -64.472 MeV | 59.93079 u | 0.9988464 u | 9.1462 MeV | 0.565612 u | 526.864 MeV | 8.7811 MeV |
| $^{62}$Ni | 28 | 34 | -66.7461 MeV | 61.928345 u | 0.9988443 u | 9.1481 MeV | 0.585383 u | 545.281 MeV | 8.7948 MeV |

$^{56}$Fe has the lowest nucleon-specific mass of the four nuclides listed in this table, but this does not imply it is the strongest bound atom per hadron, unless the choice of beginning hadrons is completely free. Iron releases the largest energy if any 56 nucleons are allowed to build a nuclide—changing one to another if necessary, The highest binding energy per hadron, with the hadrons starting as the same number of protons $Z$ and total nucleons $A$ as in the bound nucleus, is $^{62}$Ni. Thus, the true absolute value of the total binding energy of a nucleus depends on what we are allowed to construct the nucleus out of. If all nuclei of mass number $A$ were to be allowed to be constructed of $A$ neutrons, then $^{56}$Fe would release the most energy per nucleon, since it has a larger fraction of protons than $^{62}$Ni. However, if nucleons are required to be constructed of only the same number of protons and neutrons that they contain, then nickel-62 is the most tightly bound nucleus, per nucleon.

| Some light nuclides resp. atoms | | | | | | | | | |
|---|---|---|---|---|---|---|---|---|---|
| **nuclide** | ***Z*** | ***N*** | **mass excess** | **total mass** | **total mass / *A*** | **total binding energy / *A*** | **mass defect** | **binding energy** | **binding energy / *A*** |
| n | 0 | 1 | 8.0716 MeV | 1.008665 u | 1.008665 u | 0.0000 MeV | 0 u | 0 MeV | 0 MeV |
| $^{1}$H | 1 | 0 | 7.2890 MeV | 1.007825 u | 1.007825 u | 0.7826 MeV | 0.0000000146 u | 0.0000136 MeV | 13.6 eV |
| $^{2}$H | 1 | 1 | 13.13572 MeV | 2.014102 u | 1.007051 u | 1.50346 MeV | 0.002388 u | 2.22452 MeV | 1.11226 MeV |
| $^{3}$H | 1 | 2 | 14.9498 MeV | 3.016049 u | 1.005350 u | 3.08815 MeV | 0.0091058 u | 8.4820 MeV | 2.8273 MeV |
| $^{3}$He | 2 | 1 | 14.9312 MeV | 3.016029 u | 1.005343 u | 3.09433 MeV | 0.0082857 u | 7.7181 MeV | 2.5727 MeV |

In the table above it can be seen that the decay of a neutron, as well as the transformation of tritium into helium-3, releases energy; hence, it manifests a stronger bound new state when measured against the mass of an equal number of neutrons (and also a lighter state per number of total hadrons). Such reactions are not driven by changes in binding energies as calculated from previously fixed $N$ and $Z$ numbers of neutrons and protons, but rather in decreases in the total mass of the nuclide/per nucleon, with the reaction. (Note that the Binding Energy given above for hydrogen-1 is the atomic binding energy, not the nuclear binding energy which would be zero.)

## Nuclear Binding Energy

It has been seen from several observations that some nuclei are extremely stable and they don't transform to other nuclei by any radioactive decay process.* On the other hand, some nuclei are not stable and they undergo radioactive decay process and transform naturally to other nuclei. The solution to this problem lies in the understanding of nuclear binding energy. It has been observed that when a nucleus is formed from its constituent particles, its mass is less than the sum of the masses of the individual particles. This difference in mass (Δm) also known as mass defect gets converted into an amount of energy $\Delta E = (\Delta m)c^2$. This energy is known as the binding energy (B.E.) of the nucleus. Therefore, one can write the following relation:

$$B.E. = (Zm_p + Nm_n)C^2 - M_N C^2$$

where,

Z = no. of protons

N = no. of neutrons

$m_p$ = rest mass of a proton

$m_n$ = rest mass of a neutron

$M_N$ = nuclear rest mass

But the nuclear rest mass is $M_N = M_a - Zm_e$, where $M_a$ represents the atomic mass corresponding to the specified nuclear mass, Zme is the total mass of the orbiting electrons. The binding energy of the electrons is neglected, as it is quite small compared to the nuclear binding energy.

$\therefore$ $B.E. = (Zm_p + Nm_n)C^2 - (M_a - Zm_e)C^2$

Again proton mass $m_p = m_H - m_e$, where $m_H$ is the atomic mass of hydrogen atom.

$\therefore$ $B.E. = Z(m_H - m_e)C^2 + NM_nC^2 - (M_aC^2 - Zm_e)C^2$

$= Z(m_H + N m_e)C^2 + M_aC^2$

In atomic mass units the binding energy can be expressed as:

$B.E. = (Zm_H + Nm_n) - M_a$

The nuclei having positive binding energies are stable and in order to disrupt the nucleus into its constituents, energy is to be supplied from outside. However, the nuclei whose binding energy is negative are unstable and get disintegrated by themselves. The maximum binding energy per nucleon corresponds to A = 56 (Fe) and 62 (Ni). The binding energy/nucleon for $^{56}Fe$ and $^{62}Ni$ are 8.790 MeV/nucleon and 8.795 MeV/nucleon respectively. In fact, $^{62}Ni$ is the most stable nucleus in the periodic table. However, stars are generally enriched with $^{56}Fe$. Why? You will find its answer in Shurtleff and Derringh's article. The light and very heavy nuclei contain less bound nucleons. Thus the source of energy production in fusion of light nuclei or fission of very heavy nuclei can be a source of energy. They are basically the basis of nuclear energy production in reactors and weapons.

## References


- Pospelov, M.; Ritz, A. (2005). "Electric dipole moments as probes of new physics". Annals of Physics. 318: 119–169. Bibcode:2005AnPhy.318..119P. arXiv:hep-ph/0504231. doi:10.1016/j.aop.2005.04.002
- Fewell, M. P. (1995). "The atomic nuclide with the highest mean binding energy". American Journal of Physics. 63 (7): 653–658. Bibcode:1995AmJPh..63..653F. doi:10.1119/1.17828
- Bjorken, J.D.; Drell, S.D. (1964). Relativistic Quantum Mechanics. New York: McGraw-Hill. pp. 241–246. ISBN 0070054932
- Alvarez, L. W; Bloch, F. (1940). "A quantitative determination of the neutron magnetic moment in absolute nuclear magnetons". Physical Review. 57: 111–122. Bibcode:1940PhRv...57..111A. doi:10.1103/physrev.57.111
- Meshik, A. P. (November 2005). "The Workings of an Ancient Nuclear Reactor". Scientific American. 293: 82–91. doi:10.1038/scientificamerican1105-82. Retrieved 2014-01-04

- Hausser, O. (1981). "Nuclear Moments". In Lerner, R. G.; Trigg, G. L. Encyclopedia of Physics. Reading, Massachusetts: Addison-Wesley Publishing Company. pp. 679–680. ISBN 0201043130
- Hughes, D. J.; Burgy, M. T. (1949). "Reflection and polarization of neutrons by magnetized mirrors" (PDF). Phys. Rev. 76: 1413–1414. doi:10.1103/PhysRev.76.1413
- Kincade, Kathy (2 February 2015). "Pinpointing the magnetic moments of nuclear matter". Phys.org. Retrieved May 8, 2015
- Beringer, J.; et al. (Particle Data Group) (2012). "Review of Particle Physics, 2013 partial update" (PDF). Phys. Rev. D. 86: 010001. Bibcode:2012PhRvD..86a0001B. doi:10.1103/PhysRevD.86.010001. Retrieved May 8, 2015
- Schreckenbach, K. (2013). "Physics of the Neutron". In Stock, R. Encyclopedia of Nuclear Physics and its Applications. Weinheim, Germany: Wiley-VCH Verlag GmbH & Co. pp. 321–354. ISBN 978-3-527-40742-2
- Sherwood, J. E.; Stephenson, T. E.; Bernstein, S. (1954). "Stern-Gerlach experiment on polarized neutrons". Phys. Rev. 96: 1546. doi:10.1103/PhysRev.96.1546
- Esterman, I.; Stern, O. (1934). "Magnetic moment of the deuton". Physical Review. 45: 761(A109). Bibcode:1934PhRv...45..739S. doi:10.1103/PhysRev.45.739. Retrieved May 9, 2015
- John S. Rigden (1987). Rabi, Scientist and Citizen. New York: Basic Books, Inc. pp. 99–114. ISBN 9780674004351. Retrieved May 9, 2015
- Oku, T.; Suzuki, J.; et al. (2007). "Highly polarized cold neutron beam obtained by using a quadrupole magnet". Physica B. 397: 188–191. Bibcode:2007PhyB..397..188O. doi:10.1016/j.physb.2007.02.055
- Rabi, I.I.; Kellogg, J.M.; Zacharias, J.R. (1934). "The magnetic moment of the proton". Physical Review. 46: 157–163. Bibcode:1934PhRv...46..157R. doi:10.1103/physrev.46.157. Retrieved May 9, 2015
- Jacobsen, Neil E. (2007). NMR spectroscopy explained. Hoboken, New Jersey: Wiley-Interscience. ISBN 9780471730965. Retrieved May 8, 2015
- Schwinger, J. (1948). "On Quantum-Electrodynamics and the Magnetic Moment of the Electron". Physical Review. 73 (4): 416–417. Bibcode:1948PhRv...73..416S. doi:10.1103/PhysRev.73.416
- Breit, G.; Rabi, I.I. (1934). "On the interpretation of present values of nuclear moments". Physical Review. 46: 230–231. Bibcode:1934PhRv...46..230B. doi:10.1103/physrev.46.230. Retrieved May 9, 2015
- B. D. Cullity, C. D. Graham (2008). Introduction to Magnetic Materials (2 ed.). Hoboken, New Jersey: Wiley-IEEE Press. p. 103. ISBN 0-471-47741-9. Retrieved May 8, 2015
- Gell, Y.; Lichtenberg, D. B. (1969). "Quark model and the magnetic moments of proton and neutron". Il Nuovo Cimento A. Series 10. 61: 27–40. Bibcode:1969NCimA..61...27G. doi:10.1007/BF02760010
- Drell, S.; Pagels, H.R. (1965). "Anomalous Magnetic Moment of the Electron, Muon, and Nucleon". Physical Review. 140: B397–B407. Bibcode:1965PhRv..140..397D. doi:10.1103/PhysRev.140.B397. Retrieved May 10, 2015

2

# Nuclear Structures and Models

Understanding the structure of the nucleus is central to nuclear physics. To understand nuclear structure, various models are applied, out of which the liquid drop model is the simplest that has proven to best comprehend the structure of a nucleus. However, some features of nucleus structure cannot be understood using this method for which other models like nuclear shell model, interacting Boson model, etc. are used. This section discusses in detail the theories and methodologies related to models of the nucleus.

## Nuclear Structure

Understanding the structure of the atomic nucleus is one of the central challenges in nuclear physics.

### Models

### The Liquid Drop Model

The liquid drop model is one of the first models of nuclear structure, proposed by Carl Friedrich von Weizsäcker in 1935. It describes the nucleus as a semiclassical fluid made up of neutrons and protons, with an internal repulsive electrostatic force proportional to the number of protons. The quantum mechanical nature of these particles appears via the Pauli exclusion principle, which states that no two nucleons of the same kind can be at the same state. Thus the fluid is actually what is known as a Fermi liquid. In this model, the binding energy of a nucleus with $Z$ protons and $N$ neutrons is given by

$$E_B = a_V A - a_S A^{2/3} - a_C \frac{Z^2}{A^{1/3}} - a_A \frac{(N-Z)^2}{A} - \delta(A,Z)$$

where $A = Z + N$ is the total number of nucleons. The terms proportional to $A$ and $A^{2/3}$ represent the volume and surface energy of the liquid drop, the term proportional to $Z^2$ represents the electrostatic energy, the term proportional to $(N-Z)^2$ represents the Pauli exclusion principle and the last term $\delta(A,Z)$ is the pairing term, which lowers the energy for even numbers of protons or neutrons. The coefficients $a_V, a_S, a_C, a_A$ and the strength of the pairing term may be estimated theoretically, or fit to data. This simple model reproduces the main features of the binding energy of nuclei.

The assumption of nucleus as a drop of Fermi liquid is still widely used in the form of Finite Range Droplet Model (FRDM), due to the possible good reproduction of nuclear binding energy on the whole chart, with the necessary accuracy for predictions of unknown nuclei.

## The Shell Model

The expression "shell model" is ambiguous in that it refers to two different eras in the state of the art. It was previously used to describe the existence of nucleon shells in the nucleus according to an approach closer to what is now called mean field theory. Nowadays, it refers to a formalism analogous to the configuration interaction formalism used in quantum chemistry. We shall introduce the latter here.

## Introduction to the Shell Concept

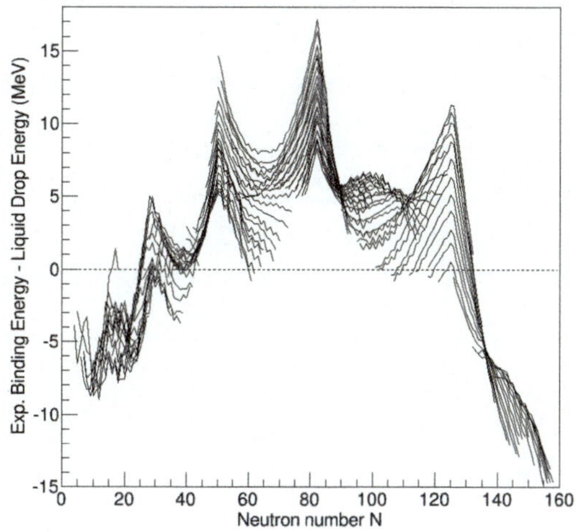


Difference between experimental binding energies and the liquid drop model prediction as a function of neutron number for Z>7

Systematic measurements of the binding energy of atomic nuclei show systematic deviations with respect to those estimated from the liquid drop model. In particular, some nuclei having certain values for the number of protons and/or neutrons are bound more tightly together than predicted by the liquid drop model. These nuclei are called singly/doubly magic. This observation led scientists to assume the existence of a shell structure of nucleons (protons and neutrons) within the nucleus, like that of electrons within atoms.

Indeed, nucleons are quantum objects. Strictly speaking, one should not speak of energies of individual nucleons, because they are all correlated with each other. However, as an approximation one may envision an average nucleus, within which nucleons propagate individually. Owing to their quantum character, they may only occupy *discrete* energy levels. These levels are by no means uniformly distributed; some intervals of energy are crowded, and some are empty, generating a gap in possible energies. A shell is such a set of levels separated from the other ones by a wide empty gap.

The energy levels are found by solving the Schrödinger equation for a single nucleon moving in the average potential generated by all other nucleons. Each level may be occupied by a nucleon, or empty. Some levels accommodate several different quantum states with the same energy; they are said to be *degenerate*. This occurs in particular if the average nucleus has some symmetry.

The concept of shells allows one to understand why some nuclei are bound more tightly than others. This is because two nucleons of the same kind cannot be in the same state (Pauli exclusion principle). So the lowest-energy state of the nucleus is one where nucleons fill all energy levels from the bottom up to some level. A nucleus with full shells is exceptionally stable, as will be explained.

As with electrons in the electron shell model, protons in the outermost shell are relatively loosely bound to the nucleus if there are only few protons in that shell, because they are farthest from the center of the nucleus. Therefore, nuclei which have a full outer proton shell will be more tightly bound and have a higher binding energy than other nuclei with a similar total number of protons. All this is also true for neutrons.

Furthermore, the energy needed to excite the nucleus (i.e. moving a nucleon to a higher, previously unoccupied level) is exceptionally high in such nuclei. Whenever this unoccupied level is the next after a full shell, the only way to excite the nucleus is to raise one nucleon *across the gap*, thus spending a large amount of energy. Otherwise, if the highest occupied energy level lies in a partly filled shell, much less energy is required to raise a nucleon to a higher state in the same shell.

Some evolution of the shell structure observed in stable nuclei is expected away from the valley of stability. For example, observations of unstable isotopes have shown shifting and even a reordering of the single particle levels of which the shell structure is composed. This is sometimes observed as the creation of an island of inversion or in the reduction of excitation energy gaps above the traditional magic numbers.

## Basic Hypotheses

Some basic hypotheses are made in order to give a precise conceptual framework to the shell model:

- The atomic nucleus is a quantum $n$-body system.
- The internal motion of nucleons within the nucleus is non-relativistic, and their behavior is governed by the Schrödinger equation.
- Nucleons are considered to be pointlike, without any internal structure.

## Brief Description of the Formalism

The general process used in the shell model calculations is the following. First a Hamiltonian for the nucleus is defined. Usually, for computational practicality, only one- and two-body terms are taken into account in this definition. The interaction is an effective theory : it contains free parameters which have to be fitted with experimental data.

The next step consists in defining a basis of single-particle states, i.e. a set of wavefunctions describing all possible nucleon states. Most of the time, this basis is obtained via a Hartree–Fock computation. With this set of one-particle states, Slater determinants are built, that is, wavefunctions for $Z$ proton variables or $N$ neutron variables, which are antisymmetrized products of single-particle wavefunctions (antisymmetrized meaning that under exchange of variables for any pair of nucleons, the wavefunction only changes sign).

In principle, the number of quantum states available for a single nucleon at a finite energy is finite, say $n$. The number of nucleons in the nucleus must be smaller than the number of available states, otherwise the nucleus cannot hold all of its nucleons. There are thus several ways to choose $Z$ (or $N$) states among the $n$ possible. In combinatorial mathematics, the number of choices of $Z$ objects among $n$ is the binomial coefficient $C^Z_n$. If $n$ is much larger than $Z$ (or $N$), this increases roughly like $n^Z$. Practically, this number becomes so large that every computation is impossible for $A=N+Z$ larger than 8.

To obviate this difficulty, the space of possible single-particle states is divided into a core and a valence shell, by analogy with chemistry. The core is a set of single-particles which are assumed to be inactive, in the sense that they are the well bound lowest-energy states, and that there is no need to reexamine their situation. They do not appear in the Slater determinants, contrary to the states in the valence space, which is the space of all single-particle states *not in the core*, but possibly to be considered in the choice of the build of the ($Z$-) $N$-body wavefunction. The set of all possible Slater determinants in the valence space defines a basis for ($Z$-) $N$-body states.

The last step consists in computing the matrix of the Hamiltonian within this basis, and to diagonalize it. In spite of the reduction of the dimension of the basis owing to the fixation of the core, the matrices to be diagonalized reach easily dimensions of the order of $10^9$, and demand specific diagonalization techniques.

The shell model calculations give in general an excellent fit with experimental data. They depend however strongly on two main factors:

- The way to divide the single-particle space into core and valence.
- The effective nucleon–nucleon interaction.

## Mean Field Theories

## The Independent-particle Model

The interaction between nucleons, which is a consequence of strong interactions and binds the nucleons within the nucleus, exhibits the peculiar behaviour of having a finite range: it vanishes when the distance between two nucleons becomes too large; it is attractive at medium range, and repulsive at very small range. This last property correlates with the Pauli exclusion principle according to which two fermions (nucleons are fermions) cannot be in the same quantum state. This results in a very large mean free path predicted for a nucleon within the nucleus.

The main idea of the Independent Particle approach is that a nucleon moves inside a certain potential well (which keeps it bound to the nucleus) independently from the other nucleons. This amounts to replacing a $N$-body problem ($N$ particles interacting) by $N$ single-body problems. This essential simplification of the problem is the cornerstone of mean field theories. These are also widely used in atomic physics, where electrons move in a mean field due to the central nucleus and the electron cloud itself.

The independent particle model and mean field theories (we shall see that there exist several variants) have a great success in describing the properties of the nucleus starting from an effective interaction or an effective potential, thus are a basic part of atomic nucleus theory. One should also notice that they are modular enough, in that it is quite easy to extend the model to introduce effects such as nuclear pairing, or collective motions of the nucleon like rotation, or vibration, adding the corresponding energy terms in the formalism. This implies that in many representation the mean field is only a starting point for a more complete description which introduces correlations reproducing properties like collective excitations and nucleon transfer.

## Nuclear Potential and Effective Interaction

A large part of the practical difficulties met in mean field theories is the definition (or calculation) of the potential of the mean field itself. One can very roughly distinguish between two approaches:

- The phenomenological approach is a parameterization of the nuclear potential by an appropriate mathematical function. Historically, this procedure was applied with the greatest success by Sven Gösta Nilsson, who used as a potential a (deformed) harmonic oscillator potential. The most recent parameterizations are based on more realistic functions, which account more accurately for scattering experiments, for example. In particular the form known as the Woods Saxon potential can be mentioned.

- The self-consistent or Hartree–Fock approach aims to deduce mathematically the nuclear potential from an effective nucleon–nucleon interaction. This technique implies a resolution of the Schrödinger equation in an iterative fashion, starting from an ansatz wavefunction and improving it variationally, since the potential depends there upon the wavefunctions to be determined. The latter are written as Slater determinants.

In the case of the Hartree–Fock approaches, the trouble is not to find the mathematical function which describes best the nuclear potential, but that which describes best the nucleon–nucleon interaction. Indeed, in contrast with atomic physics where the interaction is known (it is the Coulomb interaction), the nucleon–nucleon interaction within the nucleus is not known analytically.

There are two main reasons for this fact. First, the strong interaction acts essentially among the quarks forming the nucleons. The nucleon–nucleon interaction in vacuum is a mere *consequence* of the quark–quark interaction. While the latter is well understood in the framework of the Standard Model at high energies, it is much more complicated in low energies due to color confinement and asymptotic freedom. Thus there is yet no fundamental theory allowing one to deduce the nucleon–nucleon interaction from the quark–quark interaction. Furthermore, even if this problem were solved, there would remain a large difference between the ideal (and conceptually simpler) case of two nucleons interacting in vacuum, and that of these nucleons interacting in the nuclear matter. To go further, it was necessary to invent the concept of effective interaction. The latter is basically a mathematical function with several arbitrary parameters, which are adjusted to agree with experimental data.

Most modern interaction are zero-range so they act only when the two nucleons are in contact, as introduced by Tony Skyrme.

## The Self-consistent Approaches of the Hartree–Fock Type

In the Hartree–Fock approach of the $n$-body problem, the starting point is a Hamiltonian containing $n$ kinetic energy terms, and potential terms. As mentioned before, one of the mean field theory hypotheses is that only the two-body interaction is to be taken into account. The potential term of the Hamiltonian represents all possible two-body interactions in the set of $n$ fermions. It is the first hypothesis.

The second step consists in assuming that the wavefunction of the system can be written as a Slater determinant of one-particle spin-orbitals. This statement is the mathematical translation of the independent-particle model. This is the second hypothesis.

There remains now to determine the components of this Slater determinant, that is, the individual wavefunctions of the nucleons. To this end, it is assumed that the total wavefunction (the Slater determinant) is such that the energy is minimum. This is the third hypothesis.

Technically, it means that one must compute the mean value of the (known) two-body Hamiltonian on the (unknown) Slater determinant, and impose that its mathematical variation vanishes. This leads to a set of equations where the unknowns are the individual wavefunctions: the Hartree–Fock equations. Solving these equations gives the wavefunctions and individual energy levels of nucleons, and so the total energy of the nucleus and its wavefunction.

This short account of the Hartree–Fock method explains why it is called also the variational approach. At the beginning of the calculation, the total energy is a "function of the individual wavefunctions" (a so-called functional), and everything is then made in order to optimize the choice of these wavefunctions so that the functional has a minimum – hopefully absolute, and not only local. To be more precise, there should be mentioned that the energy is a functional of the density, defined as the sum of the individual squared wavefunctions. Let us note also that the Hartree–Fock method is also used in atomic physics and condensed matter physics as Density Functional Theory, DFT.

The process of solving the Hartree–Fock equations can only be iterative, since these are in fact a Schrödinger equation in which the potential depends on the density, that is, precisely on the wavefunctions to be determined. Practically, the algorithm is started with a set of individual grossly reasonable wavefunctions (in general the eigenfunctions of a harmonic oscillator). These allow to compute the density, and therefrom the Hartree–Fock potential. Once this done, the Schrödinger equation is solved anew, and so on. The calculation stops – convergence is reached – when the difference among wavefunctions, or energy levels, for two successive iterations is less than a fixed value. Then the mean field potential is completely determined, and the Hartree–Fock equations become standard Schrödinger equations. The corresponding Hamiltonian is then called the Hartree–Fock Hamiltonian.

## The Relativistic Mean Field Approaches

Born first in the 1970s with the works of John Dirk Walecka on quantum hadrodynamics, the relativistic models of the nucleus were sharpened up towards the end of the 1980s by P. Ring and coworkers. The starting point of these approaches is the relativistic quantum field theory. In this context, the nucleon interactions occur via the exchange of virtual particles called mesons. The idea is, in a first step, to build a Lagrangian containing these interaction terms. Second, by an application of the least action principle, one gets a set of equations of motion. The real particles (here the nucleons) obey the Dirac equation, whilst the virtual ones (here the mesons) obey the Klein–Gordon equations.

In view of the non-perturbative nature of strong interaction, and also in view of the fact that the exact potential form of this interaction between groups of nucleons is relatively badly known, the use of such an approach in the case of atomic nuclei requires drastic approximations. The main simplification consists in replacing in the equations all field terms (which are operators in the mathematical sense) by their mean value (which are functions). In this way, one gets a system of coupled integro-differential equations, which can be solved numerically, if not analytically.

## The Interacting Boson Model

The interacting boson model (IBM) is a model in nuclear physics in which nucleons are represented as pairs, each of them acting as a boson particle, with integral spin of 0, 2 or 4. This makes calculations feasible for larger nuclei. There are several branches of this model - in one of them (IBM-1) one can group all types of nucleons in pairs, in others (for instance - IBM-2) one considers protons and neutrons in pairs separately.

## Spontaneous Breaking of Symmetry in Nuclear Physics

One of the focal points of all physics is symmetry. The nucleon–nucleon interaction and all effective interactions used in practice have certain symmetries. They are invariant by translation (changing the frame of reference so that directions are not altered), by rotation (turning the frame of reference around some axis), or parity (changing the sense of axes) in the sense that the interaction does not change under any of these operations. Nevertheless, in the Hartree–Fock approach, solutions which are not invariant under such a symmetry can appear. One speaks then of spontaneous symmetry breaking.

Qualitatively, these spontaneous symmetry breakings can be explained in the following way : in the mean field theory, the nucleus is described as a set of independent particles. Most additional correlations among nucleons which do not enter the mean field are neglected. They can appear however by a breaking of the symmetry of the mean field Hamiltonian, which is only approximate. If the density used to start the iterations of the Hartree–Fock process breaks certain symmetries, the final Hartree–Fock Hamiltonian may break these symmetries, if it is advantageous to keep these broken from the point of view of the total energy.

It may also converge towards a symmetric solution. In any case, if the final solution breaks the symmetry, for example, the rotational symmetry, so that the nucleus appears not to be spherical, but elliptic, all configurations deduced from this deformed nucleus by a rotation are just as good solutions for the Hartree–Fock problem. The ground state of the nucleus is then *degenerate*.

A similar phenomenon happens with the nuclear pairing, which violates the conservation of the number of baryons.

## Extensions of the Mean Field Theories

## Nuclear Pairing Phenomenon

The most common extension to mean field theory is the nuclear pairing. Nuclei with an even number of nucleons are systematically more bound than those with an odd one. This implies that each nucleon binds with another one to form a pair, consequently the system cannot be described as independent particles subjected to a common a mean field. When the nucleus has an even number of protons and neutrons, each one of them finds a partner. To excite such a system, one must at least use such an energy as to break a pair. Conversely, in the case of odd number of protons or neutrons, there exists an unpaired nucleon, which needs less energy to be excited.

This phenomenon is closely analogous to that of Type 1 superconductivity in solid state physics. The first theoretical description of nuclear pairing was proposed at the end of the 1950s by Aage

Bohr, Ben Mottelson, and David Pines (which contributed to the reception of the Nobel Prize in Physics in 1975 by Bohr and Mottelson). It was close to the BCS theory of Bardeen, Cooper and Schrieffer, which accounts for metal superconductivity. Theoretically, the pairing phenomenon as described by the BCS theory combines with the mean field theory: nucleons are both subject to the mean field potential and to the pairing interaction.

The Hartree–Fock–Bogolyubov (HFB) method is a more sophisticated approach, enabling one to consider the pairing and mean field interactions consistently on equal footing. HFB is now the de facto standard in the mean field treatment of nuclear systems.

### Symmetry Restoration

Peculiarity of mean field methods is the calculation of nuclear property by explicit symmetry breaking. The calculation of the mean field with self-consistent methods (e.g. Hartree-Fock), breaks rotational symmetry, and the calculation of pairing property breaks particle-number.

Several techniques for symmetry restoration by projecting on good quantum numbers have been developed.

### Particle Vibrations Coupling

Mean field methods (eventually considering symmetry restoration) are a good approximation for the ground state of the system, even postulating a system of independent particles. Higher-order corrections consider the fact that the particles interact together by the means of correlation. These correlations can be introduced taking into account the coupling of independent particle degrees of freedom, low-energy collective excitation of systems with even number of protons and neutrons.

In these way excited states can be reproduced by the means of Random Phase Approximation (RPA), and eventually consistently calculating also corrections to the ground state (e.g. by the means of Nuclear Field Theory ).

## Semi-empirical Mass Formula

In nuclear physics, the semi-empirical mass formula (SEMF) (sometimes also called Weizsäcker's formula, or the Bethe–Weizsäcker formula, or the Bethe–Weizsäcker mass formula to distinguish it from the Bethe–Weizsäcker process) is used to approximate the mass and various other properties of an atomic nucleus from its number of protons and neutrons. As the name suggests, it is based partly on theory and partly on empirical measurements. The theory is based on the liquid drop model proposed by George Gamow, which can account for most of the terms in the formula and gives rough estimates for the values of the coefficients. It was first formulated in 1935 by German physicist Carl Friedrich von Weizsäcker, and although refinements have been made to the coefficients over the years, the structure of the formula remains the same today.

The SEMF gives a good approximation for atomic masses and several other effects, but does not explain the appearance of magic numbers of protons and neutrons, and the extra binding-energy and measure of stability that are associated with these numbers of nucleons.

## The Liquid Drop Model and its Analysis

The liquid drop model in nuclear physics treats the nucleus as a drop of incompressible nuclear fluid of very high density. It was first proposed by George Gamow and then developed by Niels Bohr and John Archibald Wheeler. The nucleus is made of nucleons (protons and neutrons), which are held together by the nuclear force (a residual effect of the strong force). This is very similar to the structure of spherical liquid drop made of microscopic molecules. This is a crude model that does not explain all the properties of the nucleus, but does explain the spherical shape of most nuclei. It also helps to predict the nuclear binding energy and to assess how much is available for consumption.

Mathematical analysis of the theory delivers an equation which attempts to predict the binding energy of a nucleus in terms of the numbers of protons and neutrons it contains. This equation has five terms on its right hand side. These correspond to the cohesive binding of all the nucleons by the nuclear force, a surface energy term, the electrostatic (Coulomb) mutual repulsion of the protons, an asymmetry term (derivable from the protons and neutrons occupying independent quantum momentum states) and a pairing term (partly derivable from the protons and neutrons occupying independent quantum spin states).

If we consider the sum of the following five types of energies, then the picture of a nucleus as a drop of incompressible liquid roughly accounts for the observed variation of binding energy of the nucleus:

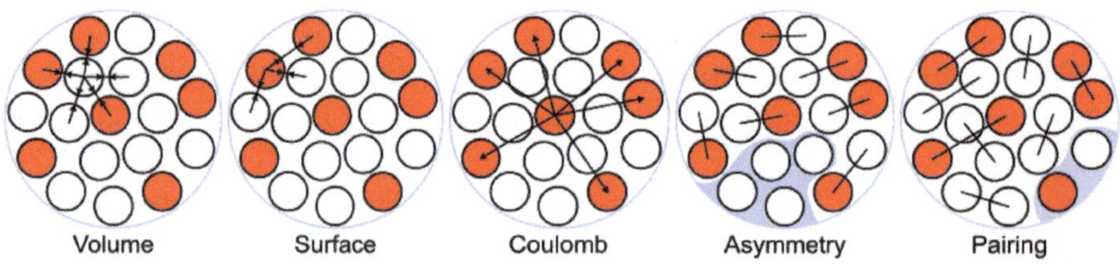


Volume energy. When an assembly of nucleons of the same size is packed together into the smallest volume, each interior nucleon has a certain number of other nucleons in contact with it. So, this nuclear energy is proportional to the volume.

Surface energy. A nucleon at the surface of a nucleus interacts with fewer other nucleons than one in the interior of the nucleus and hence its binding energy is less. This surface energy term takes that into account and is therefore negative and is proportional to the surface area.

Coulomb Energy. The electric repulsion between each pair of protons in a nucleus contributes toward decreasing its binding energy.

Asymmetry energy (also called Pauli Energy). An energy associated with the Pauli exclusion principle. Were it not for the Coulomb energy, the most stable form of nuclear matter would have the same number of neutrons as protons, since unequal numbers of neutrons and protons imply filling higher energy levels for one type of particle, while leaving lower energy levels vacant for the other type.

Pairing energy. An energy which is a correction term that arises from the tendency of proton pairs and neutron pairs to occur. An even number of particles is more stable than an odd number.

## The Formula

In the following formulae, let $A$ be the total number of nucleons, $Z$ the number of protons, and $N$ the number of neutrons, so that $A = Z + N$.

The mass of an atomic nucleus is given by

$$m = Zm_p + Nm_n - \frac{E_B}{c^2}$$

where $m_p$ and $m_n$ are the rest mass of a proton and a neutron, respectively, and $E_B$ is the binding energy of the nucleus. The semi-empirical mass formula states that the binding energy will take the following form:

$$E_B = a_V A - a_S A^{2/3} - a_C \frac{Z^2}{A^{1/3}} - a_A \frac{(A-2Z)^2}{A} \pm \delta(A,Z)$$

Each of the terms in this formula has a theoretical basis, as will be explained below. The coefficients $a_V$, $a_S$, $a_C$, $a_A$ and a coefficient that appears in the formula for $\delta(A,Z)$ are determined empirically.

## Terms

## Volume Term

The term $a_V A$ is known as the *volume term*. The volume of the nucleus is proportional to $A$, so this term is proportional to the volume, hence the name.

The basis for this term is the strong nuclear force. The strong force affects both protons and neutrons, and as expected, this term is independent of $Z$. Because the number of pairs that can be taken from $A$ particles is $\frac{A(A-1)}{2}$, one might expect a term proportional to $A^2$. However, the strong force has a very limited range, and a given nucleon may only interact strongly with its nearest neighbors and next nearest neighbors. Therefore, the number of pairs of particles that actually interact is roughly proportional to $A$, giving the volume term its form.

The coefficient $a_V$ is smaller than the binding energy of the nucleons to their neighbours $E_b$, which is of order of 40 MeV. This is because the larger the number of nucleons in the nucleus, the larger their kinetic energy is, due to the Pauli exclusion principle. If one treats the nucleus as a Fermi ball of $A$ nucleons, with equal numbers of protons and neutrons, then the total kinetic energy is $\frac{3}{5}A\varepsilon_F$, with $\varepsilon_F$ the Fermi energy which is estimated as 28 MeV. Thus the expected value of $a_V$ in this model is $E_b - \frac{3}{5}\varepsilon_F \sim 17\text{ MeV}$, not far from the measured value.

## Surface Term

The term $a_S A^{2/3}$ is known as the *surface term*. This term, also based on the strong force, is a correction to the volume term.

The volume term suggests that each nucleon interacts with a constant number of nucleons, independent of $A$. While this is very nearly true for nucleons deep within the nucleus, those nucleons on the surface of the nucleus have fewer nearest neighbors, justifying this correction. This can also be thought of as a surface tension term, and indeed a similar mechanism creates surface tension in liquids.

If the volume of the nucleus is proportional to $A$, then the radius should be proportional to $A^{1/3}$ and the surface area to $A^{2/3}$. This explains why the surface term is proportional to $A^{2/3}$. It can also be deduced that $A^{2/3}$ should have a similar order of magnitude as $a_V$.

## Coulomb Term

The term $a_C \dfrac{Z(Z-1)}{A^{1/3}}$ or $a_C \dfrac{Z^2}{A^{1/3}}$ is known as the *Coulomb* or *electrostatic term.*

The basis for this term is the electrostatic repulsion between protons. To a very rough approximation, the nucleus can be considered a sphere of uniform charge density. The potential energy of such a charge distribution can be shown to be

$$E = \frac{3}{5}\left(\frac{1}{4\pi\varepsilon_0}\right)\frac{Q^2}{R}$$

where $Q$ is the total charge and $R$ is the radius of the sphere. Identifying $Q$ with $Ze$, and noting as above that the radius is proportional to $A^{1/3}$, we get close to the form of the Coulomb term. However, because electrostatic repulsion will only exist for more than one proton, $Z^2$ becomes $Z(Z-1)$. The value of $a_C$ can be approximately calculated using the equation above:

Empirical nuclear radius:

$$R \approx r_0 A^{\frac{1}{3}}.$$

Quantum charge integers:

$$Q = Ze$$

$$Z^2 \approx Z(Z-1).$$

Integration by substitution:

$$E = \frac{3}{5}\left(\frac{1}{4\pi\varepsilon_0}\right)\frac{Q^2}{R} = \frac{3}{5}\left(\frac{1}{4\pi\varepsilon_0}\right)\frac{(Ze)^2}{(r_0 A^{\frac{1}{3}})} = \frac{3e^2 Z^2}{20\pi\varepsilon_0 r_0 A^{\frac{1}{3}}} \approx \frac{3e^2 Z(Z-1)}{20\pi\varepsilon_0 r_0 A^{\frac{1}{3}}} = a_C \frac{Z(Z-1)}{A^{1/3}}$$

Potential energy of charge distribution:

$$E = \frac{3e^2 Z(Z-1)}{20\pi\varepsilon_0 r_0 A^{\frac{1}{3}}}$$

Electrostatic Coulomb constant:

$$a_C = \frac{3e^2}{20\pi\varepsilon_0 r_0}$$

The value of $a_C$ using the fine structure constant:

$$a_C = \frac{3}{5}\left(\frac{\hbar c\alpha}{r_0}\right) = \frac{3}{5}\left(\frac{R_P}{r_0}\right)\alpha m_p c^2$$

where $\alpha$ is the fine structure constant and $r_0 A^{1/3}$ is the radius of a nucleus, giving $r_0$ to be approximately 1.25 femtometers. $R_P$ is the proton Compton radius and $m_p$ the proton mass. This gives $a_C$ an approximate theoretical value of 0.691 MeV, not far from the measured value.

$$a_C = 0.691 \text{ MeV}$$

## Asymmetry Term

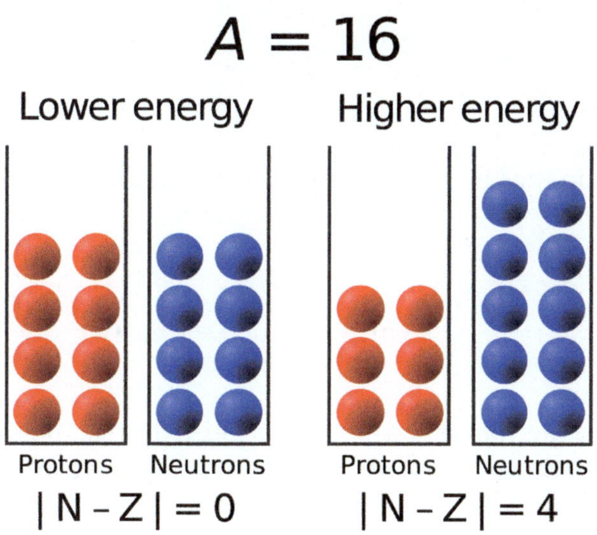


The term $a_A \frac{(A-2Z)^2}{A}$ or $4a_A \frac{((A/2)-Z)^2}{A}$ is known as the *asymmetry term*. Note that as $A = N + Z$, the parenthesized expression can be rewritten as $(N-Z)$. The form $(A-2Z)$ is used to keep the dependence on *A* explicit, as it will be important for a number of uses of the formula.

The theoretical justification for this term is more complex. The Pauli exclusion principle states that no two identical fermions can occupy exactly the same quantum state in an atom. At a given energy level, there are only finitely many quantum states available for particles. What this means in the nucleus is that as more particles are "added", these particles must occupy higher energy levels, increasing the total energy of the nucleus (and decreasing the binding energy). Note that this effect is not based on any of the fundamental forces (gravitational, electromagnetic, etc.), only the Pauli exclusion principle.

Protons and neutrons, being distinct types of particles, occupy different quantum states. One can think of two different "pools" of states, one for protons and one for neutrons. Now, for example, if there are significantly more neutrons than protons in a nucleus, some of the neutrons will be higher in energy than the available states in the proton pool. If we could move some particles from the neutron pool to the proton pool, in other words change some neutrons into protons, we would significantly decrease the energy. The imbalance between the number of protons and neutrons causes the energy to be higher than it needs to be, *for a given number of nucleons*. This is the basis for the asymmetry term.

The actual form of the asymmetry term can again be derived by modelling the nucleus as a Fermi ball of protons and neutrons. Its total kinetic energy is

$$E_k = \frac{3}{5}(N_p \varepsilon_{Fp} + N_n \varepsilon_{Fn})$$

where $N_p$, $N_n$ are the numbers of protons and neutrons and $\varepsilon_{Fp}$, $\varepsilon_{Fn}$ are their Fermi energies. Since the latter are proportional to $N_p^{2/3}$ and $N_n^{2/3}$, respectively, one gets

$E_k = C(N_p^{5/3} + N_n^{5/3})$ for some constant *C*.

The leading expansion in the difference $N_n - N_p$ is then

$$E_k = \frac{C}{2^{2/3}}\left( (N_p + N_n)^{5/3} + \frac{5}{9}\frac{(N_n - N_p)^2}{(N_p + N_n)^{1/3}} \right) + O((N_n - N_p)^2).$$

At the zeroth order expansion the kinetic energy is just the Fermi energy $\varepsilon_F \equiv \varepsilon_{Fp} = \varepsilon_{Fn}$ multiplied by $\frac{3}{5}(N_p + N_n)^{2/3}$. Thus we get

$$E_k = \frac{3}{5}\varepsilon_F (N_p + N_n) + \frac{1}{3}\varepsilon_F \frac{(N_n - N_p)^2}{(N_p + N_n)} + O((N_n - N_p)^4) = \frac{3}{5}\varepsilon_F A + \frac{1}{3}\varepsilon_F \frac{(A-2Z)^2}{A} + O((A-2Z)^4).$$

The first term contributes to the volume term in the semi-empirical mass formula, and the second term is minus the asymmetry term (remember the kinetic energy contributes to the total binding energy with a *negative* sign).

$\varepsilon_F$ is 38 MeV, so calculating $a_A$ from the equation above, we get only half the measured value. The discrepancy is explained by our model not being accurate: nucleons in fact interact with each other, and are not spread evenly across the nucleus. For example, in the shell model, a proton and a neutron with overlapping wavefunctions will have a greater strong interaction between them and stronger binding energy. This makes it energetically favourable (i.e. having lower energy) for protons and neutrons to have the same quantum numbers (other than isospin), and thus increase the energy cost of asymmetry between them.

One can also understand the asymmetry term intuitively, as follows. It should be dependent on the absolute difference $|N - Z|$, and the form $(A - 2Z)^2$ is simple and differentiable, which is important for certain applications of the formula. In addition, small differences between *Z* and *N* do not have a high energy cost. The *A* in the denominator reflects the fact that a given difference $|N - Z|$ is less significant for larger values of *A*.

## Pairing Term

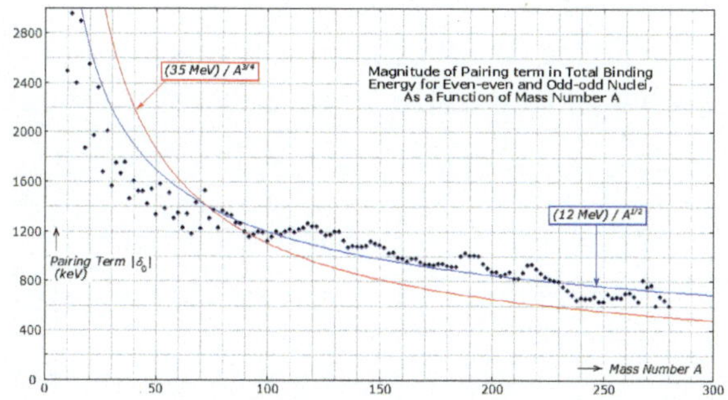


Magnitude of the pairing term in the total binding energy for even-even and odd-odd nuclei, as a function of mass number. Two fits are shown (blue and red line). The pairing term (positive for even-even and negative for odd-odd nuclei) was derived from the binding energy data in: G. Audi et al., 'The AME2012 atomic mass evaluation', in Chinese Physics C 36 (2012/12) pp. 1287–1602.

The term $\delta(A,Z)$ is known as the *pairing term* (possibly also known as the pairwise interaction). This term captures the effect of spin-coupling. It is given by:

$$\delta(A,Z) = \begin{cases} +\delta_0 & Z, N \text{ even } (A \text{ even}) \\ 0 & A \text{ odd} \\ -\delta_0 & Z, N \text{ odd } (A \text{ even}) \end{cases}$$

where $\delta_0$ is found empirically to have a value of about 1000 keV, slowly decreasing with mass number $A$. The dependence on mass number is commonly parametrized as

$$\delta_0 = \frac{a_P}{A^{k_P}}.$$

The value of the exponent $k_P$ is determined from experimental binding energy data. In the past its value was often assumed to be −3/4, but modern experimental data indicate that a value of −1/2 is nearer the mark:

$$\delta_0 = \frac{a_P}{A^{1/2}} \text{ or } \delta_0 = \frac{a_P{}'}{A^{3/4}}.$$

Due to the Pauli exclusion principle the nucleus would have a lower energy if the number of protons with spin up were equal to the number of protons with spin down. This is also true for neutrons. Only if both $Z$ and $N$ are even can both protons and neutrons have equal numbers of spin up and spin down particles. This is a similar effect to the asymmetry term.

The factor $A^{-k_P}$ is not easily explained theoretically. The Fermi ball calculation we have used above, based on the liquid drop model but neglecting interactions, will give an $A^{-1}$ dependence, as in the asymmetry term. This means that the actual effect for large nuclei will be larger than expected by that model. This should be explained by the interactions between nucleons; For example, in the shell model, two protons with the same quantum numbers (other than spin) will have completely overlapping wavefunctions and will thus have greater strong interaction between them and stronger binding energy. This makes it energetically favourable (i.e. having lower energy) for protons to pair in pairs of opposite spin. The same is true for neutrons.

## Calculating the Coefficients

The coefficients are calculated by fitting to experimentally measured masses of nuclei. Their values can vary depending on how they are fitted to the data. Several examples are as shown below, with units of megaelectronvolts.

| | Least-squares fit (1) | Least-squares fit (2) | Rohlf | Wapstra |
|---|---|---|---|---|
| $a_V$ | 15.8 | 15.76 | 15.75 | 14.1 |
| $a_S$ | 18.3 | 17.81 | 17.8 | 13 |
| $a_C$ | 0.714 | 0.711 | 0.711 | 0.595 |
| $a_A$ | 23.2 | 23.702 | 23.7 | 19 |
| $a_P$ | 12 | 34 | 11.18 | 33.5 |
| **exponent of $A$ in $\delta_0$** | −1/2 | −3/4 | −1/2 | −3/4 |
| $\delta_0$ **(even-even)** | $\frac{+12}{A^{1/2}}$ | $\frac{+34}{A^{3/4}}$ | $\frac{+11.18}{A^{1/2}}$ | $\frac{+33.5}{A^{3/4}}$ |
| $\delta_0$ **(odd-odd)** | $\frac{-12}{A^{1/2}}$ | $\frac{-34}{A^{3/4}}$ | $\frac{-33.5}{A^{3/4}}$ | $\frac{-33.5}{A^{3/4}}$ |
| $\delta_0$ **(even-odd, odd-even)** | 0 | 0 | 0 | 0 |

The semi-empirical mass formula provides a good fit to heavier nuclei, and a poor fit to very light nuclei, especially $^4$He. This is because the formula does not consider the internal shell structure of the nucleus. For light nuclei, it is usually better to use a model that takes this structure into account.

## Examples for Consequences of the Formula

By maximizing $E_b(A,Z)$ with respect to $Z$, we find the best neutrons to protons ratio $N/Z$ for a given atomic weight $A$. We get

$$N/Z \approx 1 + \frac{a_C}{2a_A} A^{2/3}.$$

This is roughly 1 for light nuclei, but for heavy nuclei the ratio grows in good agreement with nature.

By substituting the above value of $Z$ back into $E_b$ one obtains the binding energy as a function of the atomic weight, $E_b(A)$. Maximizing $E_b(A)/A$ with respect to $A$ gives the nucleus which is most strongly bound, i.e. most stable. The value we get is $A$ = 63 (copper), close to the measured values of $A$ = 62 (nickel) and $A$ = 58 (iron).

# The Liquid Drop Model of Nucleus

To understand nuclear properties various theoretical models have been proposed by various nuclear physicists. The simplest model to understand some of the basic properties is 'Liquid drop model'.

## The Liquid Drop Model of Nucleus

Weizsacher in 1935 recognized some experimental evidences and found resemblance of nucleus with a liquid drop. Later on, in 1939 Bohr and Wheeler employed a liquid drop model of the nucleus in their theory of nuclear fission. Considerable information about nuclear behaviour across the periodic table is found in a plot between neutron numbers against proton numbers for various stable nuclei, called Segre chart as depicted in the figure.

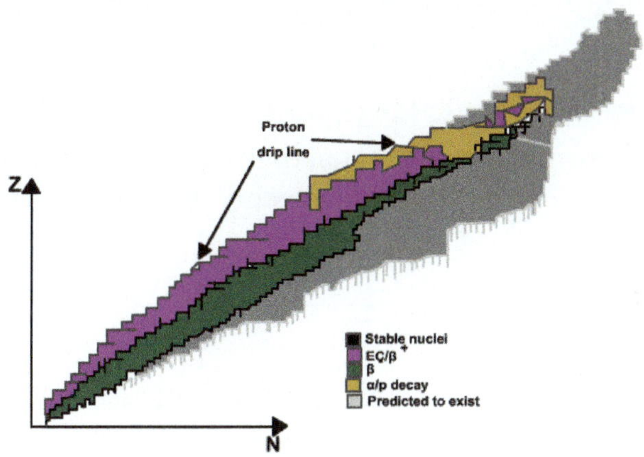


Segre chart of stable nuclides

This figure illustrates that the points representing stable nuclei goes away from the line N/Z = 1 for low Z towards the direction of higher number of neutrons as one takes heavier elements. For A = 238, N/Z = 1.6. We will see later that this trend is ascribed to an increase in Coulomb repulsion energy among protons with increase in Z. To counteract this, heavy nuclei tend to be made up of more neutrons than protons. The curve depicted in figure is known as stability curve and the isotopes lying above and below this curve are radioactive. They decay by emitting alpha and beta particles and reach towards stability curve. In order to understand this in detail, let us look at how binding energy of a nucleus varies with nuclear mass and what all other factors influence it.

If the mass number of a nucleus is $A$, binding energy per nucleon is defined by,

$$\frac{B.E.}{A} = \frac{[(Zm_H + NM_n) - M_a]c^2}{A}$$

When this quantity is plotted against mass number, the plot obtained is shown in the figure.

The decrease of $\frac{B.E.}{A}$ for large A is due to the coulomb repulsion between the protons, which clearly makes the nuclei increasingly less stable.

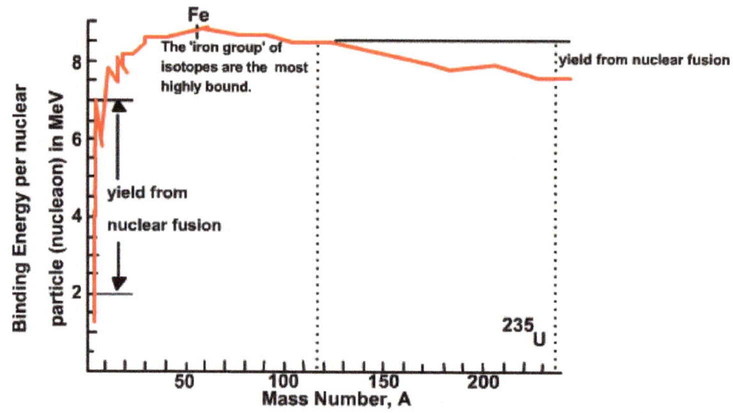


Binding energy per nucleon versus mass number of nuclei for different elements of the periodic table

From the graph it is evident that the elements in the middle of the periodic table are most stable, since they have the largest B.E./A values. The maximum stability is reached in the neighborhood of mass number 50.

Nucleons are bound inside the nucleus by a mutual interacting force known as nuclear force. Therefore, nuclear binding energy is proportional to the volume of the nucleus or to the total number of nucleons A, which contributes the volume energy term $B_1$ and can be expressed as

$B1 \propto V \propto A$

$\therefore B1 = a1A$ where $a1 > 0$

This can be better understood on the basis of liquid drop picture. For a liquid drop to evaporate, a certain amount of heat must be supplied, which is the product of heat of vaporization, mass of the material that is number of molecules times the mass of each molecule. The reason for this simple relationship is that binding energy is the sum of all interactions between molecules, and each molecule interacts only with its neighbors. This consideration is correct for a system in which the range of interactions between particles is small compared to the dimensions of the system.

However, the nucleons those are situated in the surface region of the nucleus are necessarily more weakly bound than those residing inside the nuclear interior because they have fewer immediate neighbors. The number of such nucleons is proportional to the surface area of the nucleus and therefore to $R^2$ and so is proportional to $A^{2/3}$ (because $R \propto A^{1/3}$). Thus we have

$B_2 = -\alpha_2 A^{2/3}$

Where $\alpha_2 > 0$

The sign is negative because it weakens the binding energy as it corresponds to the surface tension of a liquid drop. This effect is pronounced for lower A nuclei and hence the B.E./A decreases for lower A. $B_2$ is also referred to as surface energy term.

Now let us address the issue why at higher A, B.E./A again decreases quite significantly. Definitely, there is some sort of interaction which is trying to destabilize nuclei having higher A. The most probable reason is enhanced Coulomb repulsion between positively charged protons. The number of protons and hence the Coulomb repulsion energy increases with increase in A value. The contribution due to Coulomb energy can be calculated in the following manner: Approximately, a nucleus may be considered to have a constant charge density ρ , and for r > R (nuclear radius) there is no charge.

Therefore,

$$\rho = \frac{Ze}{\frac{4}{3}\pi R^3}$$

The work $dW$ required to bring a thin spherical shell (thickness dr) of this charge upto its radius r is obtained from the coulomb's law:

$$dW = \frac{4}{3}\pi r^3 \rho \times 4\pi r^2 dr \frac{\rho}{4\pi\varepsilon_0 r}$$

Integrating this from r = 0 to r = R, which gives total work done in assembling the sphere

$$W = \frac{3/5Z^2eh2}{4\pi\varepsilon_0 R} \text{ -}$$

Even if there were only one proton in the nucleus, we would obtain

$$W_{proton} = \frac{3/5e^2}{4\pi\varepsilon_0 R} \text{ -}$$

Now considering Z protons and from the above equations the coulomb energy term can be written as

$$-B_3 = W - W_{proton} = \frac{3/5Z(Z-1)e^2}{4\pi\varepsilon_0 R}$$

Or

$$-B_3 = -a_3 \frac{Z(Z-1)}{A^{1/3}}$$

where $a^3$ includes all the constant terms of equation.

## Liquid Drop Model: Nuclear Stability

Now, the above two analysis clearly show that the B.E./A, which is supposed to be constant (as per equation m1.14) decreases for lower and higher A values because of surface effect and Coulomb repulsion effect respectively. However, nuclear physicists have observed that the change in B.E./A value for higher A is slightly more than that expected from Coulomb repulsion effect alone. The liquid drop model does not have any explanation for this. An explanation based on single particle excitation model[1] , where nucleons are considered to occupy only certain energetic orbital has been adopted. It has been realized that as A increases an asymmetry in the nucleon number develops, i.e. number of neutrons dominate over proton numbers. What I mean is that a symmetry distribution Z = N = A/2 proves to be the energetically most favoured and any other repartition, N = (A/2) + a, Z = (A/2) - a will involve lifting particles from occupied into empty orbitals, If the average energy separation between adjacent orbitals amounts to Δ, replacing k nucleons will cost an energy loss of

$$\Delta E_{binding} = \alpha\,(\Delta\,\alpha/2)$$

and, with α = (N-Z)/2, this becomes

$$\Delta E_{binding} = (N-Z)^2\,\Delta/8$$

It has been seen by theoreticians that the average energy spacing between the single particles, Δ should vary inversely proportional to A. Therefore a correction to B. E. known as asymmetry energy term is expressed as

$B_4 = -\alpha\,4(N-Z)^2/A$

The minus sign again indicates the weakening of binding energy caused by asymmetry in N and Z.

Regarding nuclear stability another important point was noticed, i.e. even-even (even Z and even N) nuclei are most stable, even-odd and odd-even nuclei are less stable and odd-odd nuclei are most unstable. This is known as pairing effect in nuclear physics. To take into account of this pairing effect, an additional term is incorporated into the binding energy expression:

$B\,5 = +\,\delta$ for even - even nuclei

$= 0$ for even - odd nuclei and odd - even nuclei

$= -\,\delta$ for odd - odd nuclei

where δ is empirically found to be given by

$$\delta = a_5 A^{-3/4}$$

Now considering all the above mentioned terms, the B. E. of a nucleus can be expressed as:

$$B.\,E. = B_1 + B_2 + B_3 + B_4 + B_5 = a_1A - a_2A^{2/3}\ a_3\frac{Z(Z-1)}{A^{1/3}}\ -a_4(N-Z)^2/A + a_5A^{-3/4}$$

Now the B.E./A vs. A can be better depicted as shown in the figure. The effect of Coulomb repulsion and the nucleon asymmetry for lower and higher A values respectively clearly demonstrated in this figure. Overall, the resultant binding energy is to be considered, which is same as the figure given below.

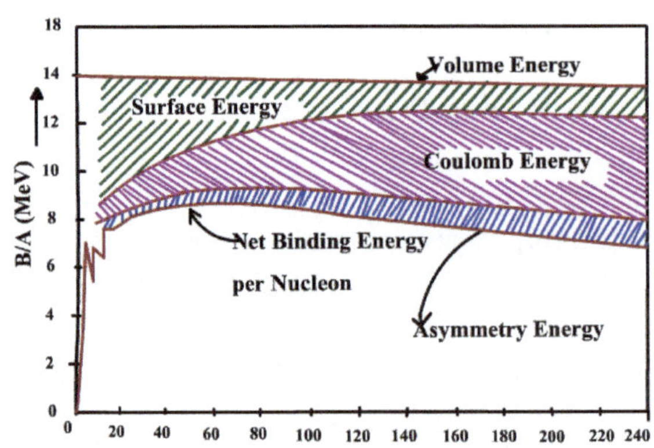


B.E./A versus A showing various energy contribution to B. E. The resultant B. E. is also clearly indicated.

Considering the above formulation of B. E. the famous nuclear physicists Weizsacker and Bethe developed a semi- empirical equation to express the mass of a nucleus as:

$$M(Z,A) = ZM_H + (A-Z)M_N - a_1A + a_2A^{2/3} + \frac{a_3Z^2}{A^{1/3}} + \frac{a_4(N-Z)^2}{A} + \Delta$$

This is popularly known as Weizsacker's semi empirical mass formula and has a great implication in nuclear physics. This equation can make a clear picture about nuclear stability, one of which is the stability against beta particle disintegration.

Using fits of known masses to this final equation one can determine the coefficients $a_1$ to $a_5$ as shown by Wapstra and Audi[4] :

$a_1$ = 15.85 MeV

$a_2$ = 18.34 MeV

$a_3$ = 0.71 MeV

$a_4$ = 23.21 MeV

$a_5$ = 12.0 MeV

The volume energy amounts to almost 16 MeV per nucleon: the various corrections than steadily brings this value towards the empirical value as observed throughout the nuclear mass region.

## Magic Number (Physics)

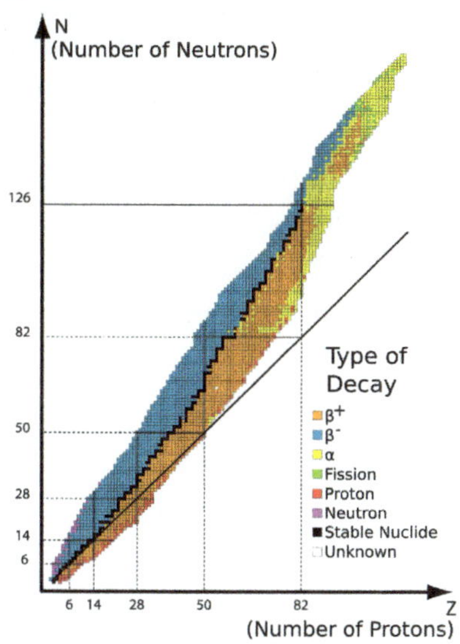


Graph of isotope stability.

In nuclear physics, a magic number is a number of nucleons (either protons or neutrons, separately) such that they are arranged into complete shells within the atomic nucleus. The seven most widely recognized magic numbers as of 2007 are 2, 8, 20, 28, 50, 82, and 126 (sequence A018226 in the OEIS). Atomic nuclei consisting of such a magic number of nucleons have a higher average binding energy per nucleon than one would expect based upon predictions such as the semi-empirical mass formula and are hence more stable against nuclear decay.

The unusual stability of isotopes having magic numbers means that transuranium elements can be created with extremely large nuclei and yet not be subject to the extremely rapid radioactive decay normally associated with high atomic numbers. Large isotopes with magic numbers of nucleons are said to exist in an island of stability. Unlike the magic numbers 2–126, which are realized in spherical nuclei, theoretical calculations predict that nuclei in the island of stability are deformed. Before this was realized, higher magic numbers, such as 184, 258, 350, and 462 (sequence A033547 in the OEIS), were predicted based on simple calculations that assumed spherical shapes: these are generated by the formula $2\left(\binom{n}{1}+\binom{n}{2}+\binom{n}{3}\right)$. It is now believed that the sequence of spherical magic numbers cannot be extended in this way. Further predicted magic numbers are 114, 122, 124, and 164 for protons as well as 184, 196, 236, and 318 for neutrons.

## Origin of the Term

According to Steven A. Moszkowski (a student of Maria Goeppert-Mayer), the term "magic number" was coined by Eugene Wigner: "Wigner, too, believed in the liquid drop model, but he recognized, from the work of Maria Mayer, the very strong evidence for the closed shells. It seemed a little like magic to him, and that is how the words 'Magic Numbers' were coined."

## Double Magic

Nuclei which have neutron number and proton (atomic) numbers each equal to one of the magic numbers are called "double magic", and are especially stable against decay. Examples of double magic isotopes include helium-4, oxygen-16, calcium-40, calcium-48, nickel-48, nickel-78, and lead-208.

Double-magic effects may allow existence of stable isotopes which otherwise would not have been expected. An example is calcium-40, with 20 neutrons and 20 protons, which is the heaviest stable isotope made of the same number of protons and neutrons. Both calcium-48 and nickel-48 are double magic because calcium-48 has 20 protons and 28 neutrons while nickel-48 has 28 protons and 20 neutrons. Calcium-48 is very neutron-rich for such a light element, but like calcium-40, it is made stable by being double magic. Nickel-48, discovered in 1999, is the most proton-rich isotope known beyond helium-3. At the other extreme, nickel-78 is also doubly magical, with 28 protons and 50 neutrons, a ratio observed only in much heavier elements apart from tritium with one proton and two neutrons (Ni-78: 28/50 = 0.56; U-238: 92/146 = 0.63).

Magic number shell effects are seen in ordinary abundances of elements: helium-4 is among the most abundant (and stable) nuclei in the universe and lead-208 is the heaviest stable nuclide.

Magic effects can keep unstable nuclides from decaying as rapidly as would otherwise be expected. For example, the nuclides tin-100 and tin-132 are examples of doubly magic isotopes of tin that are unstable, and represent endpoints beyond which stability drops off rapidly.

In December 2006 hassium-270, with 108 protons and 162 neutrons, was discovered by an international team of scientists led by the Technical University of Munich having the half-life of 22 seconds. Hassium-270 evidently forms part of an island of stability, and may even be double magic.

## Derivation

Magic numbers are typically obtained by empirical studies; if the form of the nuclear potential is known then the Schrödinger equation can be solved for the motion of nucleons and energy levels determined. Nuclear shells are said to occur when the separation between energy levels is significantly greater than the local mean separation.

In the shell model for the nucleus, magic numbers are the numbers of nucleons at which a shell is filled. For instance the magic number 8 occurs when $1s_{1/2}$, $1p_{3/2}$, $1p_{1/2}$ energy levels are filled as there is a large energy gap between the $1p_{1/2}$ and the next highest $1d_{5/2}$ energy levels.

The atomic analog to nuclear magic numbers are those numbers of electrons leading to discontinuities in the ionization energy. These occur for the noble gases helium, neon, argon, krypton, xenon, radon and oganesson. Hence, the "atomic magic numbers" are 2, 10, 18, 36, 54, 86 and 118.

In 2007, Jozsef Garai from Florida International University proposed a mathematical formula describing the periodicity of the nucleus in the periodic system based on the double tetrahedron.

In 2010, an alternative explanation of magic numbers was given in terms of symmetry considerations. Based on the fractional extension of the standard rotation group, the ground state properties (including the magic numbers) for metallic clusters and nuclei were simultaneously determined analytically. A specific potential term is not necessary in this model.

# Nuclear Shell Model

In nuclear physics and nuclear chemistry, the nuclear shell model is a model of the atomic nucleus which uses the Pauli exclusion principle to describe the structure of the nucleus in terms of energy levels. The first shell model was proposed by Dmitry Ivanenko (together with E. Gapon) in 1932. The model was developed in 1949 following independent work by several physicists, most notably Eugene Paul Wigner, Maria Goeppert Mayer and J. Hans D. Jensen, who shared the 1963 Nobel Prize in Physics for their contributions.

The shell model is partly analogous to the atomic shell model which describes the arrangement of electrons in an atom, in that a filled shell results in greater stability. When adding nucleons (protons or neutrons) to a nucleus, there are certain points where the binding energy of the next nucleon is significantly less than the last one. This observation, that there are certain magic numbers of nucleons: 2, 8, 20, 28, 50, 82, 126 which are more tightly bound than the next higher number, is the origin of the shell model.

The shells for protons and for neutrons are independent of each other. Therefore, one can have "magic nuclei" where one nucleon type or the other is at a magic number, and "doubly magic nuclei", where both are. Due to some variations in orbital filling, the upper magic numbers are 126 and, speculatively, 184 for neutrons but only 114 for protons, playing a role in the search for the so-called island of stability. Some semimagic numbers have been found, notably Z=40 giving nuclear shell filling for the various elements; 16 may also be a magic number.

In order to get these numbers, the nuclear shell model starts from an average potential with a shape something between the square well and the harmonic oscillator. To this potential a spin orbit term is added. Even so, the total perturbation does not coincide with experiment, and an empirical spin orbit coupling must be added with at least two or three different values of its coupling constant, depending on the nuclei being studied.

Nevertheless, the magic numbers of nucleons, as well as other properties, can be arrived at by approximating the model with a three-dimensional harmonic oscillator plus a spin-orbit interaction. A more realistic but also complicated potential is known as Woods Saxon potential.

## Deformed Harmonic Oscillator Approximated Model

Consider a three-dimensional harmonic oscillator. This would give, for example, in the first two levels ("*l*" is angular momentum)

| level $n$ | $l$ | $m_l$ | $m_s$ |
|---|---|---|---|
| 0 | 0 | 0 | $1/2$ |
| | | | $-1/2$ |
| 1 | 1 | 1 | $1/2$ |
| | | | $-1/2$ |
| | | 0 | $1/2$ |
| | | | $-1/2$ |
| | | −1 | $1/2$ |
| | | | $-1/2$ |

We can imagine ourselves building a nucleus by adding protons and neutrons. These will always fill the lowest available level. Thus the first two protons fill level zero, the next six protons fill level one, and so on. As with electrons in the periodic table, protons in the outermost shell will be relatively loosely bound to the nucleus if there are only few protons in that shell, because they are farthest from the center of the nucleus. Therefore nuclei which have a full outer proton shell will have a higher binding energy than other nuclei with a similar total number of protons. All this is true for neutrons as well.

This means that the magic numbers are expected to be those in which all occupied shells are full. We see that for the first two numbers we get 2 (level 0 full) and 8 (levels 0 and 1 full), in accord with experiment. However the full set of magic numbers does not turn out correctly. These can be computed as follows:

In a three-dimensional harmonic oscillator the total degeneracy at level $n$ is $\frac{(n+1)(n+2)}{2}$. Due to the spin, the degeneracy is doubled and is $(n+1)(n+2)$.

Thus the magic numbers would be

$$\sum_{n=0}^{k}(n+1)(n+2) = \frac{(k+1)(k+2)(k+3)}{3}$$

for all integer $k$. This gives the following magic numbers: 2,8,20,40,70,112..., which agree with experiment only in the first three entries. These numbers are twice the tetrahedral numbers (1,4,10,20,35,56...) from the Pascal Triangle.

In particular, the first six shells are:

- level 0: 2 states ($l$ = 0) = 2.
- level 1: 6 states ($l$ = 1) = 6.
- level 2: 2 states ($l$ = 0) + 10 states ($l$ = 2) = 12.
- level 3: 6 states ($l$ = 1) + 14 states ($l$ = 3) = 20.
- level 4: 2 states ($l$ = 0) + 10 states ($l$ = 2) + 18 states ($l$ = 4) = 30.
- level 5: 6 states ($l$ = 1) + 14 states ($l$ = 3) + 22 states ($l$ = 5) = 42.

where for every $l$ there are $2l+1$ different values of $m_l$ and 2 values of $m_s$, giving a total of $4l+2$ states for every specific level.

These numbers are twice the values of triangular numbers from the Pascal Triangle: 1,3,6,10,15,21....

## Including a Spin-orbit Interaction

We next include a spin-orbit interaction. First we have to describe the system by the quantum numbers $j$, $m_j$ and parity instead of $l$, $m_l$ and $m_s$, as in the hydrogen-like atom. Since every even level includes only even values of $l$, it includes only states of even (positive) parity; Similarly every odd level includes only states of odd (negative) parity. Thus we can ignore parity in counting states. The first six shells, described by the new quantum numbers, are

- level 0 ($n$ = 0): 2 states ($j$ = $1/2$). Even parity.
- level 1 ($n$ = 1): 2 states ($j$ = $1/2$) + 4 states ($j$ = $3/2$) = 6. Odd parity.
- level 2 ($n$ = 2): 2 states ($j$ = $1/2$) + 4 states ($j$ = $3/2$) + 6 states ($j$ = $5/2$) = 12. Even parity.
- level 3 ($n$ = 3): 2 states ($j$ = $1/2$) + 4 states ($j$ = $3/2$) + 6 states ($j$ = $5/2$) + 8 states ($j$ = $7/2$) = 20. Odd parity.
- level 4 ($n$ = 4): 2 states ($j$ = $1/2$) + 4 states ($j$ = $3/2$) + 6 states ($j$ = $5/2$) + 8 states ($j$ = $7/2$) + 10 states ($j$ = $9/2$) = 30. Even parity.
- level 5 ($n$ = 5): 2 states ($j$ = $1/2$) + 4 states ($j$ = $3/2$) + 6 states ($j$ = $5/2$) + 8 states ($j$ = $7/2$) + 10 states ($j$ = $9/2$) + 12 states ($j$ = $11/2$) = 42. Odd parity.

where for every $j$ there are $2j+1$ different states from different values of $m_j$.

Due to the spin-orbit interaction the energies of states of the same level but with different $j$ will no longer be identical. This is because in the original quantum numbers, when $\vec{s}$ is parallel to $\vec{l}$, the interaction energy is positive; and in this case $j = l + s = l + 1/2$. When $\vec{s}$ is anti-parallel to $\vec{l}$ (i.e. aligned oppositely), the interaction energy is negative, and in this case $j=l-s=l-1/2$. Furthermore, the strength of the interaction is roughly proportional to $l$.

For example, consider the states at level 4:

- The 10 states with $j$ = $9/2$ come from $l$ = 4 and $s$ parallel to $l$. Thus they have a positive spin-orbit interaction energy.

- The 8 states with $j = 7/_2$ came from $l = 4$ and $s$ anti-parallel to $l$. Thus they have a negative spin-orbit interaction energy.
- The 6 states with $j = 5/_2$ came from $l = 2$ and $s$ parallel to $l$. Thus they have a positive spin-orbit interaction energy. However its magnitude is half compared to the states with $j = 9/_2$.
- The 4 states with $j = 3/_2$ came from $l = 2$ and $s$ anti-parallel to $l$. Thus they have a negative spin-orbit interaction energy. However its magnitude is half compared to the states with $j = 7/_2$.
- The 2 states with $j = 1/_2$ came from $l = 0$ and thus have zero spin-orbit interaction energy.

## Deforming the Potential

The harmonic oscillator potential $V(r) = \mu\omega^2 r^2 / 2$ grows infinitely as the distance from the center $r$ goes to infinity. A more realistic potential, such as Woods Saxon potential, would approach a constant at this limit. One main consequence is that the average radius of nucleons' orbits would be larger in a realistic potential; This leads to a reduced term $\hbar^2 l(l+1)/2mr^2$ in the Laplace operator of the Hamiltonian. Another main difference is that orbits with high average radii, such as those with high $n$ or high $l$, will have a lower energy than in a harmonic oscillator potential. Both effects lead to a reduction in the energy levels of high $l$ orbits.

## Predicted Magic Numbers

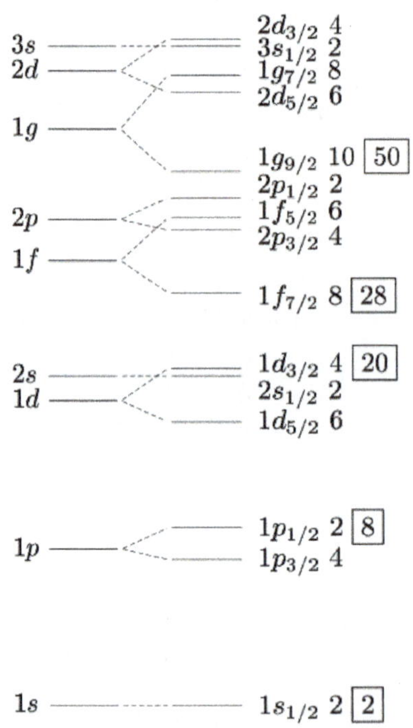


Low-lying energy levels in a single-particle shell model with an oscillator potential (with a small negative $l^2$ term) without spin-orbit (left) and with spin-orbit (right) interaction. The number to the right of a level indicates its degeneracy, $(2j+1)$. The boxed integers indicate the magic numbers.

Together with the spin-orbit interaction, and for appropriate magnitudes of both effects, one is led to the following qualitative picture: At all levels, the highest $j$ states have their energies shifted downwards, especially for high $n$ (where the highest $j$ is high). This is both due to the negative spin-orbit interaction energy and to the reduction in energy resulting from deforming the potential to a more realistic one. The second-to-highest $j$ states, on the contrary, have their energy shifted up by the first effect and down by the second effect, leading to a small overall shift. The shifts in the energy of the highest $j$ states can thus bring the energy of states of one level to be closer to the energy of states of a lower level. The "shells" of the shell model are then no longer identical to the levels denoted by $n$, and the magic numbers are changed.

We may then suppose that the highest $j$ states for $n = 3$ have an intermediate energy between the average energies of $n = 2$ and $n = 3$, and suppose that the highest $j$ states for larger $n$ (at least up to $n = 7$) have an energy closer to the average energy of $n-1$. Then we get the following shells.

- 1st shell: 2 states ($n = 0, j = 1/2$).
- 2nd shell: 6 states ($n = 1, j = 1/2$ or $3/2$).
- 3rd shell: 12 states ($n = 2, j = 1/2, 3/2$ or $5/2$).
- 4th shell: 8 states ($n = 3, j = 7/2$).
- 5th shell: 22 states ($n = 3, j = 1/2, 3/2$ or $5/2$; $n = 4, j = 9/2$).
- 6th shell: 32 states ($n = 4, j = 1/2, 3/2, 5/2$ or $7/2$; $n = 5, j = 11/2$).
- 7th shell: 44 states ($n = 5, j = 1/2, 3/2, 5/2, 7/2$ or $9/2$; $n = 6, j = 13/2$).
- 8th shell: 58 states ($n = 6, j = 1/2, 3/2, 5/2, 7/2, 9/2$ or $11/2$; $n = 7, j = 15/2$).

and so on.

The magic numbers are then

- 2
- 8=2+6
- 20=2+6+12
- 28=2+6+12+8
- 50=2+6+12+8+22
- 82=2+6+12+8+22+32
- 126=2+6+12+8+22+32+44
- 184=2+6+12+8+22+32+44+58

and so on. This gives all the observed magic numbers, and also predicts a new one (the so-called *island of stability*) at the value of 184 (for protons, the magic number 126 has not been observed yet, and more complicated theoretical considerations predict the magic number to be 114 instead).

Another way to predict magic (and semi-magic) numbers is by laying out the idealized filling order (with spin-orbit splitting but energy levels not overlapping). For consistency s is split into j = 1/2 and j = -1/2 components with 2 and 0 members respectively. Taking leftmost and rightmost total counts within sequences marked bounded by / here gives the magic and semi-magic numbers.

- s(2,0)/p(4,2) > 2,2/6,8, so (semi)magic numbers 2,2/6,8
- *d*(6,4):*s*(2,0)/*f*(8,6):*p*(4,2) > 14,18:20,20/28,34:38,40, so 14,20/28,40
- *g*(10,8):*d*(6,4):*s*(2,0)/*h*(12,10):*f*(8,6):*p*(4,2) > 50,58,64,68,70,70/82,92,100,106,110,112, so 50,70/82,112
- *i*(14,12):*g*(10,8):*d*(6,4):*s*(2,0)/*j*(16,14):*h*(12,10):*f*(8,6):*p*(4,2) > 126,138,148,156,162,166,168,168/184,198,210,220,228,234,238,240, so 126,168/184,240

The rightmost predicted magic numbers of each pair within the quartets bisected by / are double tetrahedral numbers from the Pascal Triangle: 2,8,20,40,70,112,168,240 are 2x 1,4,10,20,35,56,84,120..., and the leftmost members of the pairs differ from the rightmost by double triangular numbers: 2 – 2 = 0, 8 – 6 = 2, 20 – 14 = 6, 40 – 28 = 12, 70 – 50 = 20, 112 – 82 = 30, 168 – 126 = 42, 240 – 184 = 56, where 0,2,6,12,20,30,42,56... are 2 × 0,1,3,6,10,15,21,28....

## Other Properties of Nuclei

This model also predicts or explains with some success other properties of nuclei, in particular spin and parity of nuclei ground states, and to some extent their excited states as well. Take 17O (oxygen-17) as an example: Its nucleus has eight protons filling the three first proton "shells", eight neutrons filling the three first neutron "shells", and one extra neutron. All protons in a complete proton shell have total angular momentum zero, since their angular momenta cancel each other. The same is true for neutrons. All protons in the same level (*n*) have the same parity (either +1 or −1), and since the parity of a pair of particles is the product of their parities, an even number of protons from the same level (*n*) will have +1 parity. Thus the total angular momentum of the eight protons and the first eight neutrons is zero, and their total parity is +1. This means that the spin (i.e. angular momentum) of the nucleus, as well as its parity, are fully determined by that of the ninth neutron. This one is in the first (i.e. lowest energy) state of the 4th shell, which is a d-shell ($l$ = 2), and since $p = (-1)^l$, this gives the nucleus an overall parity of +1. This 4th d-shell has a $j = 5/2$, thus the nucleus of $^{17}{}_{8}O$ is expected to have positive parity and total angular momentum $5/2$, which indeed it has.

The rules for the ordering of the nucleus shells are similar to Hund's Rules of the atomic shells, however, unlike its use in atomic physics the completion of a shell is not signified by reaching the next *n*, as such the shell model cannot accurately predict the order of excited nuclei states, though it is very successful in predicting the ground states. The order of the first few terms are listed as follows: 1s, $1p_{3/2}$, $1p_{1/2}$, $1d_{5/2}$, 2s, $1d_{3/2}$... For further clarification on the notation refer to the article on the Russell-Saunders term symbol.

For nuclei farther from the magic numbers one must add the assumption that due to the relation between the strong nuclear force and angular momentum, protons or neutrons with the same *n* tend to form pairs of opposite angular momenta. Therefore a nucleus with an even number of protons and an even number of neutrons has 0 spin and positive parity. A nucleus with an even number of protons and an odd number of neutrons (or vice versa) has the parity of the last neutron (or proton), and the spin equal to the total angular momentum of this neutron (or proton). By "last" we mean the properties coming from the highest energy level.

In the case of a nucleus with an odd number of protons and an odd number of neutrons, one must consider the total angular momentum and parity of both the last neutron and the last proton. The nucleus parity will be a product of theirs, while the nucleus spin will be one of the possible results of the sum of their angular momenta (with other possible results being excited states of the nucleus).

The ordering of angular momentum levels within each shell is according to the principles described above - due to spin-orbit interaction, with high angular momentum states having their energies shifted downwards due to the deformation of the potential (i.e. moving from a harmonic oscillator potential to a more realistic one). For nucleon pairs, however, it is often energetically favorable to be at high angular momentum, even if its energy level for a single nucleon would be higher. This is due to the relation between angular momentum and the strong nuclear force.

Nuclear magnetic moment is partly predicted by this simple version of the shell model. The magnetic moment is calculated through $j$, $l$ and $s$ of the "last" nucleon, but nuclei are not in states of well defined $l$ and $s$. Furthermore, for odd-odd nuclei, one has to consider the two "last" nucleons, as in deuterium. Therefore one gets several possible answers for the nuclear magnetic moment, one for each possible combined $l$ and $s$ state, and the real state of the nucleus is a superposition of them. Thus the real (measured) nuclear magnetic moment is somewhere in between the possible answers.

The electric dipole of a nucleus is always zero, because its ground state has a definite parity, so its matter density ($\psi^2$, where $\psi$ is the wavefunction) is always invariant under parity. This is usually the situations with the atomic electric dipole as well.

Higher electric and magnetic multipole moments cannot be predicted by this simple version of the shell model, for the reasons similar to those in the case of deuterium.

## Including Residual Interactions

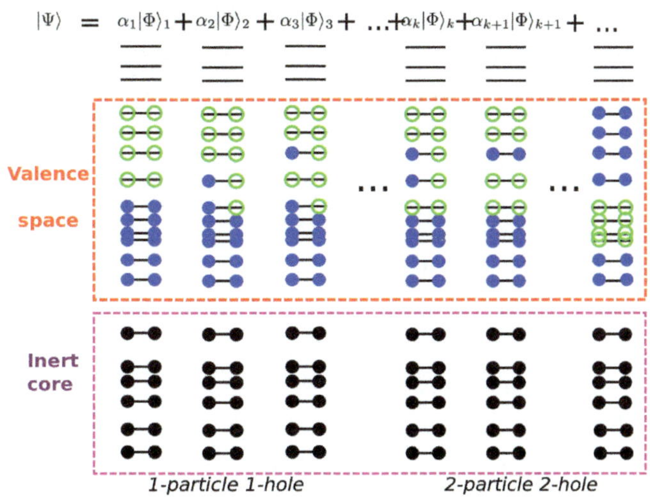


Residual interactions among valance nucleons are included by diagonalising an effective Hamiltonian in a valance space outside an inert core. As indicated, only single-particle states lying in the valance space are active in the basis used.

For nuclei having two or more valance nucleons (i.e. nucleons outside a closed shell) a residual two-body interaction must be added. This residual term comes from the part of the inter-nucleon interaction not included in the approximative average potential. Through this inclusion different shell configurations are mixed and the energy degeneracy of states corresponding to the same configuration is broken.

These residual interactions are incorporated through shell model calculations in a truncated model space (or valance space). This space is spanned by a basis of many-particle states where only single-particle states in the model space are active. The Schrödinger equation is solved in this basis, using an effective Hamiltonian specifically suited for the model space. This Hamiltonian is different from the one of free nucleons as it among other things has to compensate for excluded configurations.

One can do away with the average potential approximation entirely by extending the model space to the previously inert core and treat all single-particle states up to the model space truncation as active. This forms the basis of the no-core shell model, which is an ab initio method. It is necessary to include a three-body interaction in such calculations to achieve agreement with experiments.

## Related Models

Igal Talmi developed a method to obtain the information from experimental data and use it to calculate and predict energies which have not been measured. This method has been successfully used by many nuclear physicists and has led to deeper understanding of nuclear structure. The theory which gives a good description of these properties was developed. This description turned out to furnish the shell model basis of the elegant and successful interacting boson model.

A model derived from the nuclear shell model is the alpha particle model developed by Henry Margenau, Edward Teller, J. K. Pering, T. H. Skyrme, also sometimes called the Skyrme Model.

## Explanation of Magic Numbers--- Shell Model of a Nucleus

Whatever it may be, let me tell you that there is a resemblance between occurance of magic numbers and higher ionization potential of the noble gases (element with specific atomic number, Z), which is basically a consequence of atomic shell structure. For these atoms the electron shells are completely filled. Therefore, natuarally question arises whether nucleus too have a shell structure and if it is so, can it explain the magic numbers? It was not an easy job to frame a shell model of a nucleus similar to atomic shell model as there is no obvious central attractive force in a nucleus, like Coulomb force in an atom. Secondly, the density of nucluear matter is extremely high ($\sim 10^{17}$ kgm$^{-3}$) and hence free movement of nucleons in a nucleus like electrons in atom is difficult to imagine. However, Barlett, Guggenheimer et al. proposed an independent particle model (shell model) for the nucleus in 1933 and could able to explain lower magic numbers (2,4, 8). Later on, Maria G. Mayer in 1948 and independently Haxel et al. proposed shell structure of the nucleus and closed shell in nuclei having A corresponding to magic numbers. Their models were able to explain all magic numbers with a consideration of strong spin-orbit forces of nucleons and their coupling. Before discussing the shell structure of the nucleus given by Mayer, let us first discuss the concept of atomic shells and spin-orbit coupling. Atomic shells are basically some energy levels where electrons are allowed to stay following some quantum mechanical conditions.* As electrons move in quantized orbits, it has orbital angular momentum associated with it. Apart from orbital angular momentum it also has spin angular momentum, arises due to its spin motion, which with much simplicity can be understood as a motion about its own axis.# Not only electrons, but every elementary particles with such rotational condition have two types of angular momentum, i.e. spin angular momentum and orbital angular momentum. The first one is denoted by symbol s and the second one by symbol l. The particles like electron, proton, neutron etc. have the value of s as +1/2 or -1/2. There are several values of l, corresponding to various orbitals where the particle is moving. In case of electron, the atomic orbitals are denoted by, s, p, d, f, g, h, i, etc. and corresponding l values are 0, 1, 2, 3, 4, 5, 6 respectively. Now the l and s values of an electron (similarly, neucleons) can vectorially couple giving rise to a resultant angular momentum, j, where $j = l \pm s$, i.e. $l \pm 1/2$. For example, if we consider spectroscopic term s, then, $l = 0$, $s = 1/2$ and $j = 1/2$.

Now the next step is to consider the potential configuration within which the nucleons move inside the nucleus. The first assumption made in nuclear shell model is single particle concept with an idea that the spin, parity and moments are determined by the unpaired nucleon alone and hence essentially reducing many nucleon problems to a single particle problem. It is also assumed that each nucleon experiences a central attractive force due to the average effect of the remaining nucleons (core) in the nucleus. This is something similar to the atom where nucleus acts as a central attractive force of the electrons in the orbits. In this central field each nucleon is considered to move in a shell having specific energy and angular momentum. Another assumption of shell model is a weak inter nucleon interaction. The model of Mayer considers that the l and s values of a nucleon couples giving a distinct j value of a nucleon. Nucleons having different j values couple via j-j coupling mode, which eventually explains the origin of magic numbers.

In order to describe shell model a detail quantum mechanical calculation is required by considering some standard potential. Various potential configuration have been chosen by various authors. M. Mayer and Haxel et al. independently showed that harmonic oscillator potential or square well potential can be used to obtain reasonable results. The form of these potentials are:

(A) $V = -V_0 + 1/2M\omega^2r^2$ (linear harmonic oscillator). Where $V_0$ and $\omega$ are constants.

(B) $V = -V_0$ for $r < R$ $= 0$ for $r > R$.

Of late, more realistic potential configuration have been chosen by other Nuclear Physicists.

A detailed mathematical formulation is required to find the energy of the electrons in various shells. However, in this chapter we will just show the nucleons distribution in a closed shell and the associated magic numbers.

Shells are defined by principal quantum number n whose value runs from 0 to integer numbers. The sequence of nuclear spins with increasing n values as per standard spectroscopic notation is shown in the table.

TABLE: Filling up of nuclear shells with nucleons corresponding to magic numbers.

| Spectroscopic term | n | l | J | Number of states (2J +1) | Number of Nucleons in a shell | Magic number |
|---|---|---|---|---|---|---|
| 1s 1/2 | 0 | 0 | 1/2 | 2 | 2 | 2 |
| 1p 3/2 | 1 | 1 | 3/2 | 4 | 6 | 8 |
| 1p 3/2 | 1 | 1 | 1/2 | 2 | | |
| 1d 5/2 | 2 | 2 | 5/2 | 6 | | |
| 1d 3/2 | 2 | 2 | 3/2 | 4 | 12 | 20 |
| 1s 1/2 | 2 | 0 | 1/2 | 2 | | |
| 1f 7/2 | 3 | 3 | 7/2 | 8 | | |
| 1f 5/2 | 3 | 3 | 5/2 | 6 | 20 | 40 |
| 2p 3/2 | 3 | 1 | 3/2 | 4 | | |
| 2p 1/2 | 3 | 1 | 1/2 | 2 | | |
| 1g 9/2 | 4 | 4 | 9/2 | 10 | 10 | 50 |
| 1g 7/2 | 4 | 4 | 7/2 | 8 | 32 | 50 |
| 2d 5/2 | 4 | 2 | 2 | 6 | | |
| 2d 3/2 | 4 | 2 | 3/2 | 4 | | |
| 3s 1/2 | 4 | 0 | 1/2 | 2 | | |
| 1h 11/2 | 5 | 5 | 11/2 | 12 | | 82 |
| 1h 9/2 | 5 | 5 | 9/2 | 10 | 44 | |
| 1f 7/2 | 5 | 3 | 7/2 | 8 | | |
| 2f 5/2 | 5 | 3 | 5/2 | 6 | | |
| 3p 3/2 | 5 | 1 | 3/2 | 4 | | |
| 3p 1/2 | 5 | 1 | 1/2 | 2 | | |
| 1i 13/2 | 6 | 6 | 13/2 | 14 | | 126 |

Now it is clear from the above table that in order to understand nuclear property like magic numbers concept of shell model is must. This model has great implication to understand nuclear structure and hence nuclear spectroscopic studies.

## References


- Skyrme, T.H.R. (March 1962). "A unified field theory of mesons and baryons". Nuclear Physics. 31: 556–569. doi:10.1016/0029-5582(62)90775-7
- Garai, Jozsef (2007). "Mathematical formulas describing the sequences of the periodic table". International Journal of Quantum Chemistry. 108 (4): 667–670. Bibcode:2008IJQC..108..667G. doi:10.1002/qua.21529
- Broglia, Ricardo A.; Zelevinsky, Vladimir (2013). Fifty Years of Nuclear BCS: Pairing in Finite Systems. World Scientific. ISBN 978-981-4412-48-3

- Ozawa, A.; Kobayashi, T.; Suzuki, T.; Yoshida, K.; Tanihata, I. (2000). "New Magic Number, N=16, near the Neutron Drip Line". Physical Review Letters. 84 (24): 5493–5. Bibcode:2000PhRvL..84.5493O. PMID 10990977. doi:10.1103/PhysRevLett.84.5493
- Kratz, J. V. (5 September 2011). The Impact of Superheavy Elements on the Chemical and Physical Sciences (PDF). 4th International Conference on the Chemistry and Physics of the Transactinide Elements. Retrieved 27 August 2013
- Skyrme, T. H. R. (7 February 1961). "A Non-Linear Field Theory". Proceedings of the Royal Society A: Mathematical, Physical and Engineering Sciences. 260 (1300): 127–138. doi:10.1098/rspa.1961.0018
- Freedman, R.; Young, H. (2004). Sears and Zemanskey's University Physics with Modern Physics (11th ed.). pp. 1633–1634. ISBN 0-8053-8768-4
- Caurier, E.; Martínez-Pinedo, G.; Nowacki, F.; Poves, A.; Zuker, A. P. (2005). "The shell model as a unified view of nuclear structure". Rev. Mod. Phys. 77 (2): 427–488. doi:10.1103/RevModPhys.77.427
- Choppin, G.; Liljenzin, J.-O.; Rydberg, J. (2002). "Nuclear Mass and Stability" (PDF). Radiochemistry and Nuclear Chemistry (3rd ed.). Butterworth-Heinemann. pp. 41–57. ISBN 978-0-7506-7463-8
- Coraggio, L.; Covello, A.; Gargano, A.; Itaco, N.; Kuo, T.T.S. (2009). "Shell-model calculations and realistic effective interactions". Progress in Particle and Nuclear Physics. 62 (1): 135–182. doi:10.1016/j.ppnp.2008.06.001
- W., P. (October 23, 1999). "Twice-magic metal makes its debut - isotope of nickel". Science News. Retrieved 2006-09-29
- Barrett, B.R.; Navrátil, P.; Vary, J.P. (2013). "Ab initio no core shell model". Progress in Particle and Nuclear Physics. 69: 131–181. doi:10.1016/j.ppnp.2012.10.003

# 3

# Particle Accelerators: An Overview

To detect radiation of an electron, it is necessary to understand the means by which an electron is enhanced. There are natural ways, however, artificial means are also used as radiation detection holds a significant place in science. A renowned artificial means is the particle accelerator, which accelerates the speed of charged particles to light speed and then contains them in beams. This chapter has been carefully written to provide an easy understanding of the varied facets of particle accelerators.

## Particle Accelerator

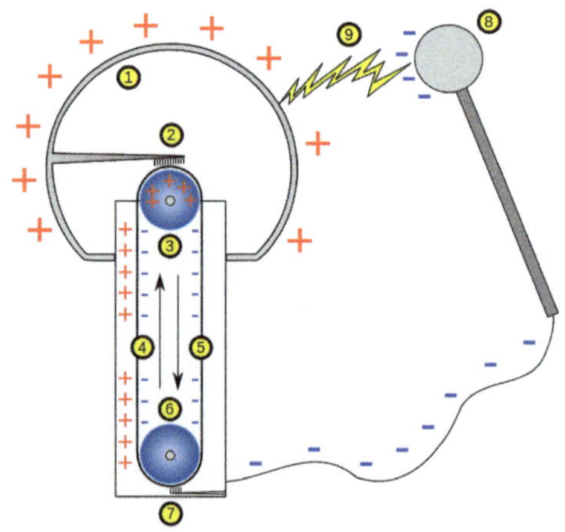

Sketch of an electrostatic Van de Graaff accelerator

A particle accelerator is a machine that uses electromagnetic fields to propel charged particles to nearly light speed and to contain them in well-defined beams. Large accelerators are used in particle physics as colliders (e.g. the LHC at CERN, KEKB at KEK in Japan, RHIC at Brookhaven National Laboratory, and Tevatron at Fermilab), or as synchrotron light sources for the study of condensed matter physics. Smaller particle accelerators are used in a wide variety of applications, including particle therapy for oncological purposes, radioisotope production for medical diagnostics, ion implanters for manufacture of semiconductors, and accelerator mass spectrometers for measurements of rare isotopes such as radiocarbon. There are currently more than 30,000 accelerators in operation around the world.

There are two basic classes of accelerators: electrostatic and electrodynamic (or electromagnetic) accelerators. *Electrostatic* accelerators use static electric fields to accelerate particles. The most common types are the Cockcroft–Walton generator and the Van de Graaff generator. A small-scale example of this class is the cathode ray tube in an ordinary old television set. The achievable kinetic energy for particles in these devices is determined by the accelerating voltage, which is limited by electrical breakdown. *Electrodynamic* or *electromagnetic* accelerators, on the other hand, use changing electromagnetic fields (either magnetic induction or oscillating radio frequency fields) to accelerate particles. Since in these types the particles can pass through the same accelerating field multiple times, the output energy is not limited by the strength of the accelerating field. This class, which was first developed in the 1920s, is the basis for most modern large-scale accelerators.

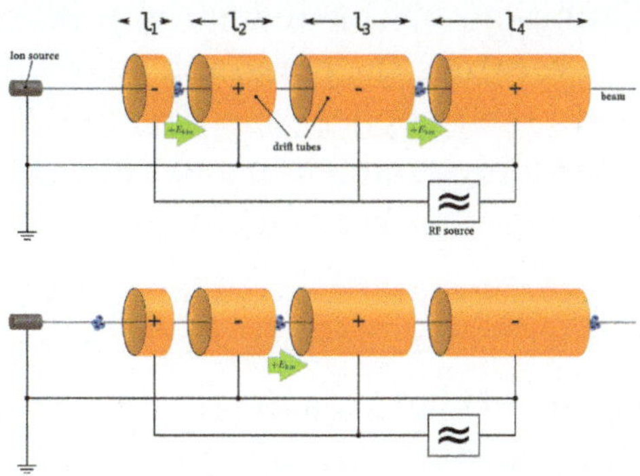


Sketch of the Ising/Widerøe linear accelerator concept, employing oscillating fields (1928)

Rolf Widerøe, Gustav Ising, Leó Szilárd, Max Steenbeck, and Ernest Lawrence are considered pioneers of this field, conceiving and building the first operational linear particle accelerator, the betatron, and the cyclotron.

Because colliders can give evidence of the structure of the subatomic world, accelerators were commonly referred to as atom smashers in the 20th century. Despite the fact that most accelerators (but not ion facilities) actually propel subatomic particles, the term persists in popular usage when referring to particle accelerators in general.

## Uses

Beamlines leading from the Van de Graaff accelerator to various experiments, in the basement of the Jussieu Campus in Paris. Breakdown of the cumulative number of industrial particle accelerators according to their applications.

The now disused Koffler particle accelerator at the Weizmann Institute, Rehovot, Israel.

Beams of high-energy particles are useful for becoming the flash andfundamental and applied research in the sciences, and also in many technical and industrial fields unrelated to fundamental research. It has been estimated that there are approximately 30,000 accelerators worldwide. Of these, only about 1% are research machines with energies above 1 GeV, while about 44% are for radiotherapy, 41% for ion implantation, 9% for industrial processing and research, and 4% for biomedical and other low-energy research. The bar graph shows the breakdown of the number of industrial accelerators according to their applications. The numbers are based on 2012 statistics available from various sources, including production and sales data published in presentations or market surveys, and data provided by a number of manufacturers.

## High-energy Physics

For the most basic inquiries into the dynamics and structure of matter, space, and time, physicists seek the simplest kinds of interactions at the highest possible energies. These typically entail particle energies of many GeV, and the interactions of the simplest kinds of particles: leptons (e.g. electrons and positrons) and quarks for the matter, or photons and gluons for the field quanta. Since isolated quarks are experimentally unavailable due to color confinement, the simplest available experiments involve the interactions of, first, leptons with each other, and second, of leptons with nucleons, which are composed of quarks and gluons. To study the collisions of quarks with each other, scientists resort to collisions of nucleons, which at high energy may be usefully considered as essentially 2-body interactions of the quarks and gluons of which they are composed. Thus elementary particle physicists tend to use machines creating beams of electrons, positrons, protons, and antiprotons, interacting with each other or with the simplest nuclei (e.g., hydrogen or deuterium) at the highest possible energies, generally hundreds of GeV or more.

The largest and highest energy particle accelerator used for elementary particle physics is the Large Hadron Collider (LHC) at CERN, operating since 2009.

## Nuclear Physics and Isotope Production

Nuclear physicists and cosmologists may use beams of bare atomic nuclei, stripped of electrons, to investigate the structure, interactions, and properties of the nuclei themselves, and of condensed matter at extremely high temperatures and densities, such as might have occurred in the first moments of the Big Bang. These investigations often involve collisions of heavy nuclei – of atoms like iron or gold – at energies of several GeV per nucleon. The largest such particle accelerator is the Relativistic Heavy Ion Collider (RHIC) at Brookhaven National Laboratory.

Particle accelerators can also produce proton beams, which can produce proton-rich medical or research isotopes as opposed to the neutron-rich ones made in fission reactors; however, recent work has shown how to make $^{99}Mo$, usually made in reactors, by accelerating isotopes of hydrogen, although this method still requires a reactor to produce tritium. An example of this type of machine is LANSCE at Los Alamos.

## Synchrotron Radiation

Besides being of fundamental interest, electrons accelerated in a magnetic field causes the high energy electrons to emit extremely bright and coherent beams of high energy photons via synchrotron radiation in the continuous spectrum, which have numerous uses in the study of atomic structure, chemistry, condensed matter physics, biology, and technology. A large number of synchrotron light sources exist worldwide. Examples in the US are SSRL and LCLS at SLAC National Accelerator Laboratory, APS at Argonne National Laboratory, ALS at Lawrence Berkeley National Laboratory, and NSLS at Brookhaven National Laboratory. The ESRF in Grenoble, France has been used to extract detailed 3-dimensional images of insects trapped in amber. Thus there is a great demand for electron accelerators of moderate (GeV) energy and high intensity.

## Low-energy Machines and Particle Therapy

Everyday examples of particle accelerators are cathode ray tubes found in television sets and X-ray generators. These low-energy accelerators use a single pair of electrodes with a DC voltage of a few thousand volts between them. In an X-ray generator, the target itself is one of the electrodes. A low-energy particle accelerator called an ion implanter is used in the manufacture of integrated circuits.

At lower energies, beams of accelerated nuclei are also used in medicine as particle therapy, for the treatment of cancer.

DC accelerator types capable of accelerating particles to speeds sufficient to cause nuclear reactions are Cockcroft-Walton generators or voltage multipliers, which convert AC to high voltage DC, or Van de Graaff generators that use static electricity carried by belts.

## Electrostatic Particle Accelerators

A Cockcroft-Walton generator (Philips, 1937), residing in Science Museum (London).

Historically, the first accelerators used simple technology of a single static high voltage to accelerate charged particles. The charged particle was accelerated through an evacuated tube with an electrode at either end, with the static potential across it. Since the particle passed only once through the potential difference, the output energy was limited to the accelerating voltage of the machine. While this method is still extremely popular today, with the electrostatic accelerators greatly out-numbering any other type, they are more suited to lower energy studies owing to the practical voltage limit of about 1 MV for air insulated machines, or 30 MV when the accelerator is operated in a tank of pressurized gas with high dielectric strength, such as sulfur hexafluoride. In a *tandem accelerator* the potential is used twice to accelerate the particles, by reversing the charge of the particles while they are inside the terminal. This is possible with the acceleration of atomic nuclei by using anions (negatively charged ions), and then passing the beam through a thin foil to strip electrons off the anions inside the high voltage terminal, converting them to cations (positively charged ions), which are accelerated again as they leave the terminal.

A 1960s single stage 2 MeV linear Van de Graaff accelerator, here opened for maintenance

The two main types of electrostatic accelerator are the Cockcroft-Walton accelerator, which uses a diode-capacitor voltage multiplier to produce high voltage, and the Van de Graaff accelerator, which uses a moving fabric belt to carry charge to the high voltage electrode. Although electrostatic accelerators accelerate particles along a straight line, the term linear accelerator is more often used for accelerators that employ oscillating rather than static electric fields.

## Electrodynamic (Electromagnetic) Particle Accelerators

Due to the high voltage ceiling imposed by electrical discharge, in order to accelerate particles to higher energies, techniques involving dynamic fields rather than static fields are used. Electrodynamic acceleration can arise from either of two mechanisms: non-resonant magnetic induction, or resonant circuits or cavities excited by oscillating RF fields. Electrodynamic accelerators can be *linear*, with particles accelerating in a straight line, or *circular*, using magnetic fields to bend particles in a roughly circular orbit.

## Magnetic Induction Accelerators

Magnetic induction accelerators accelerate particles by induction from an increasing magnetic field, as if the particles were the secondary winding in a transformer. The increasing magnetic field creates a circulating electric field which can be configured to accelerate the particles. Induction accelerators can be either linear or circular.

## Linear Induction Accelerators

Linear induction accelerators utilize ferrite-loaded, non-resonant induction cavities. Each cavity can be thought of as two large washer-shaped disks connected by an outer cylindrical tube. Between the disks is a ferrite toroid. A voltage pulse applied between the two disks causes an increasing magnetic field which inductively couples power into the charged particle beam.

The linear induction accelerator was invented by Christofilos in the 1960s. Linear induction accelerators are capable of accelerating very high beam currents (>1000 A) in a single short pulse. They have been used to generate X-rays for flash radiography (e.g. DARHT at LANL), and have been considered as particle injectors for magnetic confinement fusion and as drivers for free electron lasers.

## Betatrons

The Betatron is circular magnetic induction accelerator, invented by Donald Kerst in 1940 for accelerating electrons. The concept originates ultimately from Norwegian-German scientist Rolf Widerøe. These machines, like synchrotrons, use a donut-shaped ring magnet with a cyclically increasing B field, but accelerate the particles by induction from the increasing magnetic field, as if they were the secondary winding in a transformer, due to the changing magnetic flux through the orbit.

Achieving constant orbital radius while supplying the proper accelerating electric field requires that the magnetic flux linking the orbit be somewhat independent of the magnetic field on the orbit, bending the particles into a constant radius curve. These machines have in practice been limited by the large radiative losses suffered by the electrons moving at nearly the speed of light in a relatively small radius orbit.

## Linear Accelerators

Modern superconducting radio frequency, multicell linear accelerator component.

In a linear particle accelerator (linac), particles are accelerated in a straight line with a target of interest at one end. They are often used to provide an initial low-energy kick to particles before they are injected into circular accelerators. The longest linac in the world is the Stanford Linear Accelerator, SLAC, which is 3 km (1.9 mi) long. SLAC is an electron-positron collider.

Linear high-energy accelerators use a linear array of plates (or drift tubes) to which an alternating high-energy field is applied. As the particles approach a plate they are accelerated towards it by an opposite polarity charge applied to the plate. As they pass through a hole in the plate, the polarity is switched so that the plate now repels them and they are now accelerated by it towards the next plate. Normally a stream of "bunches" of particles are accelerated, so a carefully controlled AC voltage is applied to each plate to continuously repeat this process for each bunch.

As the particles approach the speed of light the switching rate of the electric fields becomes so high that they operate at radio frequencies, and so microwave cavities are used in higher energy machines instead of simple plates.

Linear accelerators are also widely used in medicine, for radiotherapy and radiosurgery. Medical grade linacs accelerate electrons using a klystron and a complex bending magnet arrangement which produces a beam of 6-30 MeV energy. The electrons can be used directly or they can be collided with a target to produce a beam of X-rays. The reliability, flexibility and accuracy of the radiation beam produced has largely supplanted the older use of cobalt-60 therapy as a treatment tool.

## Circular or Cyclic RF Accelerators

In the circular accelerator, particles move in a circle until they reach sufficient energy. The particle track is typically bent into a circle using electromagnets. The advantage of circular accelerators over linear accelerators (*linacs*) is that the ring topology allows continuous acceleration, as the particle can transit indefinitely. Another advantage is that a circular accelerator is smaller than a linear accelerator of comparable power (i.e. a linac would have to be extremely long to have the equivalent power of a circular accelerator).

Depending on the energy and the particle being accelerated, circular accelerators suffer a disadvantage in that the particles emit synchrotron radiation. When any charged particle is accelerated, it emits electromagnetic radiation and secondary emissions. As a particle traveling in a circle is always accelerating towards the center of the circle, it continuously radiates towards the tangent of the circle. This radiation is called synchrotron light and depends highly on the mass of the accelerating particle. For this reason, many high energy electron accelerators are linacs. Certain accelerators (synchrotrons) are however built specially for producing synchrotron light (X-rays).

Since the special theory of relativity requires that matter always travels slower than the speed of light in a vacuum, in high-energy accelerators, as the energy increases the particle speed approaches the speed of light as a limit, but never attains it. Therefore, particle physicists do not generally think in terms of speed, but rather in terms of a particle's energy or momentum, usually measured in electron volts (eV). An important principle for circular accelerators, and particle beams in general, is that the curvature of the particle trajectory is proportional to the particle charge and to the magnetic field, but inversely proportional to the (typically relativistic) momentum.

## Cyclotrons

The earliest operational circular accelerators were cyclotrons, invented in 1929 by Ernest O. Lawrence at the University of California, Berkeley. Cyclotrons have a single pair of hollow 'D'-shaped plates to accelerate the particles and a single large dipole magnet to bend their path into a circular orbit. It is a characteristic property of charged particles in a uniform and constant magnetic field B that they orbit with a constant period, at a frequency called the cyclotron frequency, so long as their speed is small compared to the speed of light *c*. This means that the accelerating D's of a cyclotron can be driven at a constant frequency by a radio frequency (RF) accelerating power source, as the beam spirals outwards continuously. The particles are injected in the centre of the magnet and are extracted at the outer edge at their maximum energy.

Lawrence's 60 inch cyclotron, with magnet poles 60 inches (5 feet, 1.5 meters) in diameter, at the University of California Lawrence Radiation Laboratory, Berkeley, in August, 1939, the most powerful accelerator in the world at the time. Glenn T. Seaborg and Edwin M. McMillan *(right)* used it to discover plutonium, neptunium and many other transuranic elements and isotopes, for which they received the 1951 Nobel Prize in chemistry.

Cyclotrons reach an energy limit because of relativistic effects whereby the particles effectively become more massive, so that their cyclotron frequency drops out of synch with the accelerating RF. Therefore, simple cyclotrons can accelerate protons only to an energy of around 15 million electron volts (15 MeV, corresponding to a speed of roughly 10% of *c*), because the protons get out of phase with the driving electric field. If accelerated further, the beam would continue to spiral outward to a larger radius but the particles would no longer gain enough speed to complete the larger circle in step with the accelerating RF. To accommodate relativistic effects the magnetic field needs to be increased to higher radii like it is done in isochronous cyclotrons. An example of an isochronous cyclotron is the PSI Ring cyclotron in Switzerland, which provides protons at the energy of 590 MeV which corresponds to roughly 80% of the speed of light. The advantage of such a cyclotron is the maximum achievable extracted proton current which is currently 2.2 mA. The energy and current correspond to 1.3 MW beam power which is the highest of any accelerator currently existing.

## Synchrocyclotrons and Isochronous Cyclotrons

A classic cyclotron can be modified to increase its energy limit. The historically first approach was the synchrocyclotron, which accelerates the particles in bunches. It uses a constant magnetic field , but reduces the accelerating field's frequency so as to keep the particles in step as they spiral outward, matching their mass-dependent cyclotron resonance frequency. This approach suffers from low average beam intensity due to the bunching, and again from the need for a huge magnet of large radius and constant field over the larger orbit demanded by high energy.

A magnet in the synchrocyclotron at the Orsay proton therapy center

The second approach to the problem of accelerating relativistic particles is the isochronous cyclotron. In such a structure, the accelerating field's frequency (and the cyclotron resonance frequency) is kept constant for all energies by shaping the magnet poles so to increase magnetic field with radius. Thus, all particles get accelerated in isochronous time intervals. Higher energy particles travel a shorter distance in each orbit than they would in a classical cyclotron, thus remaining in phase with the accelerating field. The advantage of the isochronous cyclotron is that it can deliver continuous beams of higher average intensity, which is useful for some applications. The main disadvantages are the size and cost of the large magnet needed, and the difficulty in achieving the high magnetic field values required at the outer edge of the structure.

Synchrocyclotrons have not been built since the isochronous cyclotron was developed.

## Synchrotrons

Aerial photo of the Tevatron at Fermilab, which resembles a figure eight. The main accelerator is the ring above; the one below (about half the diameter, despite appearances) is for preliminary acceleration, beam cooling and storage, etc.

To reach still higher energies, with relativistic mass approaching or exceeding the rest mass of the particles (for protons, billions of electron volts or GeV), it is necessary to use a synchrotron. This is an accelerator in which the particles are accelerated in a ring of constant radius. An immediate advantage over cyclotrons is that the magnetic field need only be present over the actual region of the particle orbits, which is much narrower than that of the ring. (The largest cyclotron built in the US had a 184-inch-diameter (4.7 m) magnet pole, whereas the diameter of synchrotrons such as the LEP and LHC is nearly 10 km. The aperture of the two beams of the LHC is of the order of a centimeter).

However, since the particle momentum increases during acceleration, it is necessary to turn up the magnetic field B in proportion to maintain constant curvature of the orbit. In consequence, synchrotrons cannot accelerate particles continuously, as cyclotrons can, but must operate cyclically, supplying particles in bunches, which are delivered to a target or an external beam in beam "spills" typically every few seconds.

Since high energy synchrotrons do most of their work on particles that are already traveling at nearly the speed of light *c*, the time to complete one orbit of the ring is nearly constant, as is the frequency of the RF cavity resonators used to drive the acceleration.

In modern synchrotrons, the beam aperture is small and the magnetic field does not cover the entire area of the particle orbit as it does for a cyclotron, so several necessary functions can be separated. Instead of one huge magnet, one has a line of hundreds of bending magnets, enclosing (or enclosed by) vacuum connecting pipes. The design of synchrotrons was revolutionized in the early 1950s with the discovery of the strong focusing concept. The focusing of the beam is handled independently by specialized quadrupole magnets, while the acceleration itself is accomplished in separate RF sections, rather similar to short linear accelerators. Also, there is no necessity that cyclic machines be circular, but rather the beam pipe may have straight sections between magnets where beams may collide, be cooled, etc. This has developed into an entire separate subject, called "beam physics" or "beam optics".

More complex modern synchrotrons such as the Tevatron, LEP, and LHC may deliver the particle bunches into storage rings of magnets with a constant magnetic field, where they can continue to orbit for long periods for experimentation or further acceleration. The highest-energy machines such as the Tevatron and LHC are actually accelerator complexes, with a cascade of specialized elements in series, including linear accelerators for initial beam creation, one or more low energy synchrotrons to reach intermediate energy, storage rings where beams can be accumulated or "cooled" (reducing the magnet aperture required and permitting tighter focusing.

Segment of an electron synchrotron at DESY

## Electron Synchrotrons

Circular electron accelerators fell somewhat out of favor for particle physics around the time that SLAC's linear particle accelerator was constructed, because their synchrotron losses were considered economically prohibitive and because their beam intensity was lower than for the unpulsed linear machines. The Cornell Electron Synchrotron, built at low cost in the late 1970s, was the first in a series of high-energy circular electron accelerators built for fundamental particle physics, the last being LEP, built at CERN, which was used from 1989 until 2000.

A large number of electron synchrotrons have been built in the past two decades, as part of synchrotron light sources that emit ultraviolet light and X rays.

## Storage Rings

For some applications, it is useful to store beams of high energy particles for some time (with modern high vacuum technology, up to many hours) without further acceleration. This is especially true for colliding beam accelerators, in which two beams moving in opposite directions are made to collide with each other, with a large gain in effective collision energy. Because relatively few collisions occur at each pass through the intersection point of the two beams, it is customary to first accelerate the beams to the desired energy, and then store them in storage rings, which are essentially synchrotron rings of magnets, with no significant RF power for acceleration.

## Synchrotron Radiation Sources

Some circular accelerators have been built to deliberately generate radiation (called synchrotron light) as X-rays also called synchrotron radiation, for example the Diamond Light Source which has been built at the Rutherford Appleton Laboratory in England or the Advanced Photon Source at Argonne National Laboratory in Illinois, USA. High-energy X-rays are useful for X-ray spectroscopy of proteins or X-ray absorption fine structure (XAFS), for example.

Synchrotron radiation is more powerfully emitted by lighter particles, so these accelerators are invariably electron accelerators. Synchrotron radiation allows for better imaging as researched and developed at SLAC's SPEAR.

## FFAG Accelerators

Fixed-Field Alternating Gradient accelerators (FFAG)s, in which a very strong radial field gradient, combined with strong focusing, allows the beam to be confined to a narrow ring, are an extension of the isochronous cyclotron idea that is lately under development. They use RF accelerating sections between the magnets, and so are isochronous for relativistic particles like electrons (which achieve essentially the speed of light at only a few MeV), but only over a limited energy range for protons and heavier particles at sub-relativistic energies. Like the isochronous cyclotrons, they achieve continuous beam operation, but without the need for a huge dipole bending magnet covering the entire radius of the orbits.

## History

Ernest Lawrence's first cyclotron was a mere 4 inches (100 mm) in diameter. Later, in 1939, he built a machine with a 60-inch diameter pole face, and planned one with a 184-inch diameter in 1942, which was, however, taken over for World War II-related work connected with uranium isotope separation; after the war it continued in service for research and medicine over many years.

The first large proton synchrotron was the Cosmotron at Brookhaven National Laboratory, which accelerated protons to about 3 GeV (1953–1968). The Bevatron at Berkeley, completed in 1954, was specifically designed to accelerate protons to sufficient energy to create antiprotons, and verify the particle-antiparticle symmetry of nature, then only theorized. The Alternating Gradient Synchrotron (AGS) at Brookhaven (1960–) was the first large synchrotron with alternating gradient, "strong focusing" magnets, which greatly reduced the required aperture of the beam, and correspondingly the size and cost of the bending magnets. The Proton Synchrotron, built at CERN (1959–), was the first major European particle accelerator and generally similar to the AGS.

The Stanford Linear Accelerator, SLAC, became operational in 1966, accelerating electrons to 30 GeV in a 3 km long waveguide, buried in a tunnel and powered by hundreds of large klystrons. It is still the largest linear accelerator in existence, and has been upgraded with the addition of storage rings and an electron-positron collider facility. It is also an X-ray and UV synchrotron photon source.

The Fermilab Tevatron has a ring with a beam path of 4 miles (6.4 km). It has received several upgrades, and has functioned as a proton-antiproton collider until it was shut down due to budget cuts on September 30, 2011. The largest circular accelerator ever built was the LEP synchrotron at CERN with a circumference 26.6 kilometers, which was an electron/positron collider. It achieved an energy of 209 GeV before it was dismantled in 2000 so that the underground tunnel could be used for the Large Hadron Collider (LHC). The LHC is a proton collider, and currently the world's largest and highest-energy accelerator, achieving 6.5 TeV energy per beam (13 TeV in total).

The aborted Superconducting Super Collider (SSC) in Texas would have had a circumference of 87 km. Construction was started in 1991, but abandoned in 1993. Very large circular accelerators are invariably built in underground tunnels a few metres wide to minimize the disruption and cost of building such a structure on the surface, and to provide shielding against intense secondary radiations that occur, which are extremely penetrating at high energies.

Current accelerators such as the Spallation Neutron Source, incorporate superconducting cryomodules. The Relativistic Heavy Ion Collider, and Large Hadron Collider also make use of superconducting magnets and RF cavity resonators to accelerate particles.

## Targets and Detectors

The output of a particle accelerator can generally be directed towards multiple lines of experiments, one at a given time, by means of a deviating electromagnet. This makes it possible to operate multiple experiments without needing to move things around or shutting down the entire accelerator beam. Except for synchrotron radiation sources, the purpose of an accelerator is to generate high-energy particles for interaction with matter.

This is usually a fixed target, such as the phosphor coating on the back of the screen in the case of a television tube; a piece of uranium in an accelerator designed as a neutron source; or a tungsten target for an X-ray generator. In a linac, the target is simply fitted to the end of the accelerator. The particle track in a cyclotron is a spiral outwards from the centre of the circular machine, so the accelerated particles emerge from a fixed point as for a linear accelerator.

For synchrotrons, the situation is more complex. Particles are accelerated to the desired energy. Then, a fast acting dipole magnet is used to switch the particles out of the circular synchrotron tube and towards the target.

A variation commonly used for particle physics research is a collider, also called a *storage ring collider*. Two circular synchrotrons are built in close proximity – usually on top of each other and using the same magnets (which are then of more complicated design to accommodate both beam tubes). Bunches of particles travel in opposite directions around the two accelerators and collide at intersections between them. This can increase the energy enormously; whereas in a fixed-target experiment the energy available to produce new particles is proportional to the square root of the beam energy, in a collider the available energy is linear.

## Higher Energies

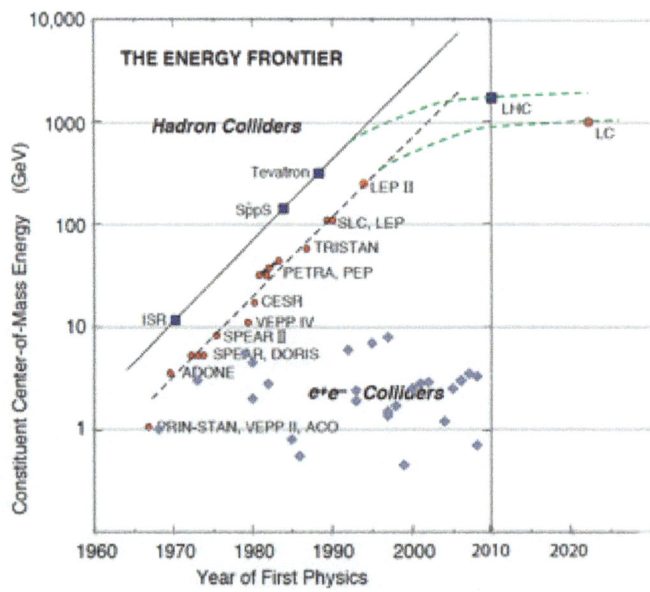

A Livingston chart depicting progress in collision energy through 2010. The LHC is the largest collision energy to date, but also represents the first break in the log-linear trend.

At present the highest energy accelerators are all circular colliders, but both hadron accelerators and electron accelerators are running into limits. Higher energy hadron and ion cyclic accelerators will require accelerator tunnels of larger physical size due to the increased beam rigidity.

For cyclic electron accelerators, a limit on practical bend radius is placed by synchrotron radiation losses and the next generation will probably be linear accelerators 10 times the current length. An example of such a next generation electron accelerator is the proposed 40 km long International Linear Collider.

It is believed that plasma wakefield acceleration in the form of electron-beam 'afterburners' and standalone laser pulsers might be able to provide dramatic increases in efficiency over RF accelerators within two to three decades. In plasma wakefield accelerators, the beam cavity is filled with a plasma (rather than vacuum). A short pulse of electrons or laser light either constitutes or immediately precedes the particles that are being accelerated. The pulse disrupts the plasma, causing the charged particles in the plasma to integrate into and move toward the rear of the bunch of particles that are being accelerated. This process transfers energy to the particle bunch, accelerating it further, and continues as long as the pulse is coherent.

Energy gradients as steep as 200 GeV/m have been achieved over millimeter-scale distances using laser pulsers and gradients approaching 1 GeV/m are being produced on the multi-centimeter-scale with electron-beam systems, in contrast to a limit of about 0.1 GeV/m for radio-frequency acceleration alone. Existing electron accelerators such as SLAC could use electron-beam afterburners to greatly increase the energy of their particle beams, at the cost of beam intensity. Electron systems in general can provide tightly collimated, reliable beams; laser systems may offer more power and compactness. Thus, plasma wakefield accelerators could be used – if technical issues can be resolved – to both increase the maximum energy of the largest accelerators and to bring high energies into university laboratories and medical centres.

Higher than 0.25 GeV/m gradients have been achieved by a dielectric laser accelerator, which may present another viable approach to building compact high-energy accelerators.

## Black Hole Production and Public Safety Concerns

In the future, the possibility of black hole production at the highest energy accelerators may arise if certain predictions of superstring theory are accurate. This and other possibilities have led to public safety concerns that have been widely reported in connection with the LHC, which began operation in 2008. The various possible dangerous scenarios have been assessed as presenting "no conceivable danger" in the latest risk assessment produced by the LHC Safety Assessment Group. If black holes are produced, it is theoretically predicted that such small black holes should evaporate extremely quickly via Bekenstein-Hawking radiation, but which is as yet experimentally unconfirmed. If colliders can produce black holes, cosmic rays (and particularly ultra-high-energy cosmic rays, UHECRs) must have been producing them for eons, but they have yet to harm anybody. It has been argued that to conserve energy and momentum, any black holes created in a collision between an UHECR and local matter would necessarily be produced moving at relativistic speed with respect to the Earth, and should escape into space, as their accretion and growth rate should be very slow, while black holes produced in colliders (with components of equal mass) would have some chance of having a velocity less than Earth escape velocity, 11.2 km per sec, and would be liable to capture and subsequent growth. Yet even on such scenarios the collisions of UHECRs with white dwarfs and neutron stars would lead to their rapid destruction, but these bodies are observed to be common astronomical objects. Thus if stable micro black holes should be produced, they must grow far too slowly to cause any noticeable macroscopic effects within the natural lifetime of the solar system.

## Accelerator Operator

An accelerator operator controls the operation of a particle accelerator used in research experiments, reviews an experiment schedule to determine experiment parameters specified by an experimenter (physicist), adjust particle beam parameters such as aspect ratio, current intensity, and position on target, communicates with and assists accelerator maintenance personnel to ensure readiness of support systems, such as vacuum, magnet power supplies and controls, low conductivity water or LCW cooling, and radiofrequency power supplies and controls, and maintains a record of accelerator related events.

# Electrostatic Nuclear Accelerator

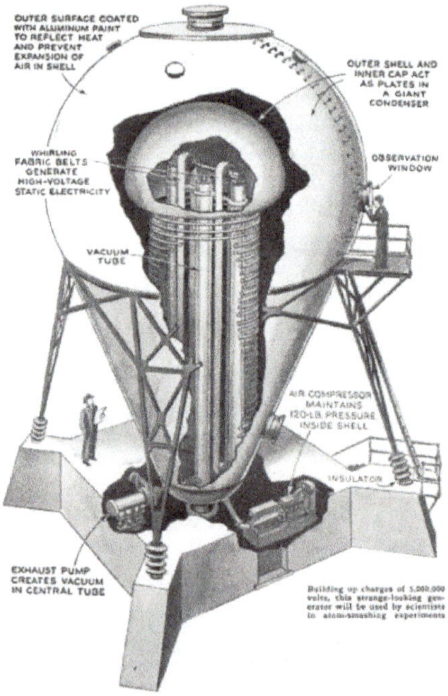


The Westinghouse Atom Smasher, an early Van de Graaff accelerator built 1937 at the Westinghouse Research Center in Forest Hills, Pennsylvania. The cutaway shows the fabric belts that carry charge up to the mushroom-shaped high voltage electrode. To improve insulation the machine was enclosed in a 65 ft. pressure vessel which was pressurized to 120 psi during operation. The high pressure air increased the voltage on the machine from 1 MV to 5 MV.

An electrostatic nuclear accelerator is one of the two main types of particle accelerators, where charged particles can be accelerated by subjection to a static high voltage potential. The static high voltage method is contrasted with the dynamic fields used in oscillating field particle accelerators. Owing to their simpler design, historically these accelerators were developed earlier. These machines are operated at lower energy than some larger oscillating field accelerators, and to the extent that the energy regime scales with the cost of these machines, in broad terms these machines are less expensive than higher energy machines, and as such they are much more common. Many universities worldwide have electrostatic accelerators for research purposes.

## Details

Although these machines accelerate atomic nuclei, the scope of application is not limited to the nuclear sciences of nuclear physics, nuclear astrophysics and nuclear chemistry. Indeed, those applications are outweighed by other uses of nuclear beams. Of the approximately 26,000 accelerators worldwide, ~44% are for radiotherapy, ~41% for ion implantation, ~9% for industrial processing and research, ~4% for biomedical and other low-energy research (less than 1% are higher energy machines).

These accelerators are being used for nuclear medicine in medical physics, sample analysis using techniques such as PIXE in the material sciences, depth profiling in solid state physics, and to a lesser extent secondary ion mass spectrometry in geologic and cosmochemical works, and even neutron beams can be made from the charged particles emerging from these accelerators to perform neutron crystallography in condensed matter physics. The principles used in electrostatic nuclear accelerators could be used to accelerate any charged particles, but particle physics operates at much higher energy regimes than these machines can achieve, and there are various better methods suited for making electron beams, so these accelerators are used for accelerating nuclei.

## Single-ended Machines

Using a high voltage terminal kept at a static potential on the order of millions of volts, charged particles can be accelerated. In simple language, an electrostatic generator is basically a giant capacitor (although lacking plates). The high voltage is achieved either using the methods of Cockcroft & Walton or Van de Graaff, with the accelerators often being named after these inventors. Van de Graaff's original design places electrons on an insulating sheet, or belt, with a metal comb, and then the sheet physically transports the immobilized electrons to the terminal. Although at high voltage, the terminal is a conductor, and there is a corresponding comb inside the conductor which can pick up the electrons off the sheet; owing to Gauss's law, there is no electric field inside a conductor, so the electrons are not repulsed by the platform once they are inside. The belt is similar in style to a conventional conveyor belt, with one major exception: it is seamless. Thus, if the belt is broken, the accelerator must be disassembled to some degree in order to replace the belt, which, owing to its constant rotation and being made typically of a rubber, is not a particularly uncommon occurrence. The practical difficulty with belts led to a different medium for physically transporting the charges: a chain of pellets. Unlike a normal chain, this one is non-conducting from one end to the other, as both insulators and conductors are used in its construction. These type of accelerators are usually called Pelletrons.

Once the platform can be electrically charged by one of the above means, some source of positive ions is placed on the platform at the end of the beam line, which is why it's called the terminal. However, as the ion source is kept at a high potential, one cannot access the ion source for control or maintenance directly. Thus, methods such as plastic rods connected to various levers inside the terminal can branch out and be toggled remotely. Omitting practical problems, if the platform is positively charged, it will repel the ions of the same electric polarity, accelerating them. As $E=qV$, where E is the emerging energy, q is the ionic charge, and V is the terminal voltage, the maximum energy of particles accelerated in this manner is practically limited by the discharge limit of the high voltage platform, about 12 MV under ambient atmospheric conditions. This limit can be increased, for example, by keeping the HV platform in a tank of an insulating gas with a higher dielectric constant than air, such as $SF_6$ which has dielectric constant roughly 2.5 times that of air. However, even in a tank of $SF_6$ the maximum attainable voltage is around 30 MV. There could be other gases with even better insulating powers, but $SF_6$ is also chemically inert and non-toxic. To increase the maximum acceleration energy further, the tandem concept was invented to use the same high voltage twice.

## Tandem Accelerators

Conventionally, positively charged ions are accelerated because this is the polarity of the atomic nucleus. However, if one wants to use the same static electric potential twice to accelerate ions, then the polarity of the ions' charge must change from anions to cations or vice versa while they are inside the conductor where they will feel no electric force. It turns out to be simple to remove, or strip, electrons from an energetic ion. One of the properties of ion interaction with matter is the exchange of electrons, which is a way the ion can lose energy by depositing it within the matter, something we should intuitively expect of a projectile shot at a solid. However, as the target becomes thinner or the projectile becomes more energetic, the amount of energy deposited in the foil becomes less and less.

Tandems locate the ion source outside the terminal, which means that accessing the ion source while the terminal is at high voltage is significantly less difficult, especially if the terminal is inside a gas tank. So then an anion beam from a sputtering ion source is injected from a relatively lower voltage platform towards the high voltage terminal. Inside the terminal, the beam impinges on a thin foil (on the order of micrograms per square centimeter), often carbon or beryllium, stripping electrons from the ion beam so that they become cations. As it is difficult to make anions of more than -1 charge state, then the energy of particles emerging from a tandem is $E=(q+1)V$, where we have added the second acceleration potential from that anion to the positive charge state q emerging from the stripper foil; we are adding these different charge signs together because we are increasing the energy of the nucleus in each phase. In this sense, we can see clearly that a tandem can double the maximum energy of a proton beam, whose maximum charge state is merely +1, but the advantage gained by a tandem has diminishing returns as we go to higher mass, as, for example, one might easily get a 6+ charge state of a silicon beam.

It is not possible to make every element into an anion easily, so it is very rare for tandems to accelerate any noble gases heavier than helium, although $KrF^-$ and $XeF^-$ have been successfully produced and accelerated with a tandem. It is not uncommon to make compounds in order to get anions, however, and $TiH_2$ might be extracted as $TiH^-$ and used to produce a proton beam, because these simple, and often weakly bound chemicals, will be broken apart at the terminal stripper foil. Anion ion beam production was a major subject of study for tandem accelerator application, and one can find recipes and yields for most elements in the Negative Ion Cookbook. Tandems can also be operated in terminal mode, where they function like a single-ended electrostatic accelerator, which is a more common and practical way to make beams of noble gases.

The name 'tandem' originates from this dual-use of the same high voltage, although tandems may also be named in the same style of conventional electrostatic accelerators based on the method of charging the terminal.

## Geometry

One trick which has to be considered with electrostatic accelerators is that usually vacuum beam lines are made of steel. However, one cannot very well connect a conducting pipe of steel from the high voltage terminal to the ground. Thus, many rings of a strong glass, like Pyrex, are assembled together in such a manner that their interface is a vacuum seal, like a copper gasket; a single long glass tube could implode under vacuum or fracture supporting its own weight. Importantly for the physics, these inter-spaced conducting rings help to make a more uniform electric field along the accelerating column. This beam line of glass rings is simply supported by compression at either end of the terminal. As the glass is non-conducting, it could be supported from the ground, but such supports near the terminal could induce a discharge of the terminal, depending on the design. Sometimes the compression is not sufficient, and the entire beam line may collapse and shatter. This idea is especially important to the design of tandems, because they naturally have longer beam lines, and the beam line must run through the terminal.

Most often electrostatic accelerators are arranged in a horizontal line. However, some tandems may have a "U" shape, and in principle the beam can be turned to any direction with a magnetic dipole at the terminal. Some electrostatic accelerators are arranged vertically, where either the ion source or, in the case of a "U" shaped vertical tandem, the terminal, is at the top of a tower. A tower arrangement can be a way to save space, and also the beam line connecting to the terminal made of glass rings can take some advantage of gravity as a natural source of compression.

## Confusion with Linear Accelerators

Electrostatic accelerators are often confused with linear accelerators simply because they can (but do not always) accelerate particles in a line. As we can see even early in their history, accelerators were named in some way referring to the method or type of acceleration. Terminal accelerators pre-date both linear accelerator technology and the nomenclature, so it would be confusing and incorrect to categorize them with a newer technology which is quite different. Linear accelerators use an array of oscillating electric fields, historically arranged in a line, but nothing would prevent a person from using magnets in between the columns of linear accelerators to form some other geometric shape. Oscillating field accelerators do not actually produce beams of particles, but rather packets of particles, unlike electrostatic accelerators which can have a beam current that is constant in time. Thus, the naming scheme for accelerators is based on the method of acceleration, or the physics, and not its geometry, which can be a point of confusion. In fact, it was the oscillating field design of the linear accelerator which inspired Lawrence to construct the cyclotron, which accelerates particles in a spiral, thus taking up a considerably smaller amount of space. A linear accelerator has more in common with a cyclotron than an electrostatic terminal accelerator.

## Understanding the Origin of the Electron Volt

The relation E=qV also indicates very simply why the electronvolt (ev) was invented for use in accelerator-based sciences, because if q is entered in integer units of the elementary charge and V in volts, the energy is given in eV; if we wanted to convert the energy into joules, we need to multiply by the elementary charge in Coulombs on both sides of the equation, yielding a very small number. Usually, the high voltage is quoted in MV and the beam energy then in MeV. With more sophisticated, higher energy machines, those working in particle physics are just accustomed to discussing mass and energy in units MeV or GeV, but the relationship of the accelerator state to the beam energy is not as simple as merely knowing the terminal voltage and the particle species accelerated.

## Linear Induction Accelerator

Linear induction accelerators utilize ferrite-loaded, non-resonant magnetic induction cavities. Each cavity can be thought of as two large washer-shaped disks connected by an outer cylindrical tube. Between the disks is a ferrite toroid. A voltage pulse applied between the two disks causes an increasing magnetic field which inductively couples power into the charged particle beam.

The linear induction accelerator was invented by Christofilos in the 1960s. Linear induction accelerators are capable of accelerating very high beam currents (>1000 A) in a single short pulse. They have been used to generate X-rays for flash radiography (e.g. DARHT at LANL), and have been considered as particle injectors for magnetic confinement fusion and as drivers for free electron lasers. A compact version of a linear induction accelerator, the dielectric wall accelerator, has been proposed as a proton accelerator for medical proton therapy.

## Linear Particle Accelerator

The linac within the Australian Synchrotron uses radio waves from a series of RF cavities at the start of the linac to accelerate the electron beam in bunches to energies of 100 MeV.

A linear particle accelerator (often shortened to linac) is a type of particle accelerator that greatly increases the kinetic energy of charged subatomic particles or ions by subjecting the charged particles to a series of oscillating electric potentials along a linear beamline. The principles for such machines were proposed by Gustav Ising in 1924, while the first machine that worked was constructed by Rolf Widerøe in 1928 at the RWTH Aachen University. Linacs have many applications: they generate X-rays and high energy electrons for medicinal purposes in radiation therapy, serve as particle injectors for higher-energy accelerators, and are used directly to achieve the highest kinetic energy for light particles (electrons and positrons) for particle physics.

The design of a linac depends on the type of particle that is being accelerated: electrons, protons or ions. Linacs range in size from a cathode ray tube (which is a type of linac) to the 3.2-kilometre-long (2.0 mi) linac at the SLAC National Accelerator Laboratory in Menlo Park, California.

## Construction and Operation

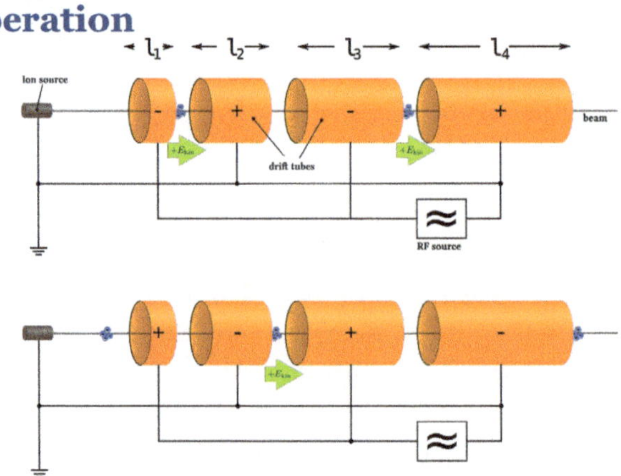


Schema of a linear accelerator

A linear particle accelerator consists of the following elements:

- The particle source. The design of the source depends on the particle that is being moved. Electrons are generated by a cold cathode, a hot cathode, a photocathode, or radio frequency (RF) ion sources. Protons are generated in an ion source, which can have many different designs. If heavier particles are to be accelerated, (e.g., uranium ions), a specialized ion source is needed.
- A high voltage source for the initial injection of particles.
- A hollow pipe vacuum chamber. The length will vary with the application. If the device is used for the production of X-rays for inspection or therapy the pipe may be only 0.5 to 1.5 meters long. If the device is to be an injector for a synchrotron it may be about ten meters long. If the device is used as the primary accelerator for nuclear particle investigations, it may be several thousand meters long.
- Within the chamber, electrically isolated cylindrical electrodes are placed, whose length varies with the distance along the pipe. The length of each electrode is determined by the frequency and power of the driving power source and the nature of the particle to be accelerated, with shorter segments near the source and longer segments near the target. The mass of the particle has a large effect on the length of the cylindrical electrodes; for example an electron is considerably lighter than a proton and so will generally require a much smaller section of cylindrical electrodes as it accelerates very quickly. Likewise, because its mass is so small, electrons have much less kinetic energy than protons at the same speed. Because of the possibility of electron emissions from highly charged surfaces, the voltages used in the accelerator have an upper limit, so this can't be as simple as just increasing voltage to match increased mass.
- One or more sources of radio frequency energy, used to energize the cylindrical electrodes. A very high power accelerator will use one source for each electrode. The sources must operate at precise power, frequency and phase appropriate to the particle type to be accelerated to obtain maximum device power.

Quadrupole magnets surrounding the linac of the Australian Synchrotron are used to help focus the electron beam

- An appropriate target. If electrons are accelerated to produce X-rays then a water cooled tungsten target is used. Various target materials are used when protons or other nuclei are accelerated, depending upon the specific investigation. For particle-to-particle collision investigations the beam may be directed to a pair of storage rings, with the particles kept within the ring by magnetic fields. The beams may then be extracted from the storage rings to create head on particle collisions.

As the particle bunch passes through the tube it is unaffected (the tube acts as a Faraday cage), while the frequency of the driving signal and the spacing of the gaps between electrodes are designed so that the maximum voltage differential appears as the particle crosses the gap. This accelerates the particle, imparting energy to it in the form of increased velocity. At speeds near the speed of light, the incremental velocity increase will be small, with the energy appearing as an increase in the mass of the particles. In portions of the accelerator where this occurs, the tubular electrode lengths will be almost constant.

- Additional magnetic or electrostatic lens elements may be included to ensure that the beam remains in the center of the pipe and its electrodes.
- Very long accelerators may maintain a precise alignment of their components through the use of servo systems guided by a laser beam.

## Advantages

The Stanford University superconducting linear accelerator, housed on campus below the Hansen Labs until 2007. This facility is separate from SLAC

Steel casting undergoing x-ray using the linear accelerator at Goodwin Steel Castings Ltd

Linacs of appropriate design are capable of accelerating heavy ions to energies exceeding those available in ring-type accelerators, which are limited by the strength of the magnetic fields required to maintain the ions on a curved path. High power linacs are also being developed for production of electrons at relativistic speeds, required since fast electrons traveling in an arc will lose energy through synchrotron radiation; this limits the maximum power that can be imparted to electrons in a synchrotron of given size. Linacs are also capable of prodigious output, producing a nearly continuous stream of particles, whereas a synchrotron will only periodically raise the particles to sufficient energy to merit a "shot" at the target. (The burst can be held or stored in the ring at energy to give the experimental electronics time to work, but the average output current is still limited.) The high density of the output makes the linac particularly attractive for use in loading storage ring facilities with particles in preparation for particle to particle collisions. The high mass output also makes the device practical for the production of antimatter particles, which are generally difficult to obtain, being only a small fraction of a target's collision products. These may then be stored and further used to study matter-antimatter annihilation.

## Medical Linacs

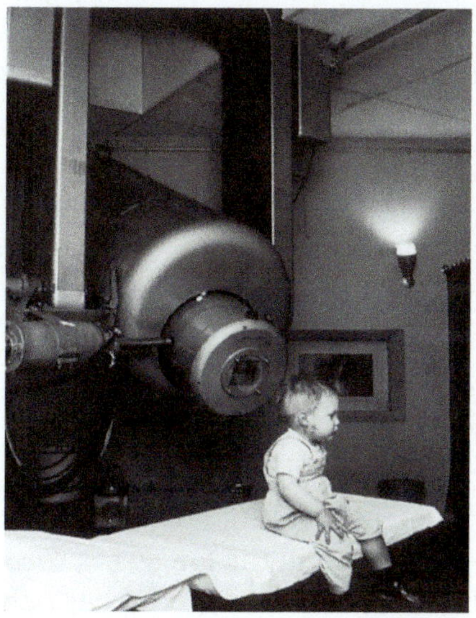

Historical image showing Gordon Isaacs, the first patient treated for retinoblastoma with linear accelerator radiation therapy (in this case an electron beam), in 1957, in the U.S. Other patients had been treated by linac for other diseases since 1953. Gordon's right eye was removed on January 11, 1957 because cancer had spread there. His left eye, however, had only a localized tumor that prompted Henry Kaplan to treat it with the electron beam.

Linac-based radiation therapy for cancer therapy began with treatment of the first patient in 1953 in London at Hammersmith Hospital, with an 8 MV machine built by Metropolitan-Vickers, as the first dedicated medical linac. A short while later in 1955, 6 MV linac therapy from a different machine was being used in the United States.

Medical grade linacs accelerate electrons using a tuned-cavity waveguide, in which the RF power creates a standing wave. Some linacs have short, vertically mounted waveguides, while higher energy machines tend to have a horizontal, longer waveguide and a bending magnet to turn the beam vertically towards the patient. Medical linacs use monoenergetic electron beams between 4 and 25 MeV, giving an X-ray output with a spectrum of energies up to and including the electron energy when the electrons are directed at a high-density (such as tungsten) target. The electrons or X-rays can be used to treat both benign and malignant disease. The LINAC produces a reliable, flexible and accurate radiation beam. The versatility of LINAC is a potential advantage over cobalt therapy as a treatment tool. In addition, the device can simply be powered off when not in use; there is no source requiring heavy shielding – although the treatment room itself requires considerable shielding of the walls, doors, ceiling etc. to prevent escape of scattered radiation. Prolonged use of high powered (>18 MeV) machines can induce a significant amount of radiation within the metal parts of the head of the machine after power to the machine has been removed (i.e. they become an active source and the necessary precautions must be observed).

## Application for Medical Isotope Development

The expected shortages with regard to Mo-99, and the technetium-99m medical isotope obtained from it, has also shed light onto linear accelerator technology to produce Mo-99 from non-enriched Uranium-235 through neutron bombardment. This would enable the medical isotope industry to manufacture this crucial isotope by a sub-critical process. The aging facilities, for example the Chalk River Laboratories in Ontario Canada, which still now produce most Mo-99 from highly enriched Uranium-235 could be replaced by this new process. In this way, the sub-critical loading of soluble uranium salts in heavy water with subsequent photo neutron bombardment and extraction of the target product, Mo-99, will be achieved.

## Disadvantages

- The device length limits the locations where one may be placed.
- A great number of driver devices and their associated power supplies are required, increasing the construction and maintenance expense of this portion.
- If the walls of the accelerating cavities are made of normally conducting material and the accelerating fields are large, the wall resistivity converts electric energy into heat quickly. On the other hand superconductors also need constant cooling to keep them below their critical temperature, and the accelerating fields are limited by quenches. Therefore, high energy accelerators such as SLAC, still the longest in the world (in its various generations), are run in short pulses, limiting the average current output and forcing the experimental detectors to handle data coming in short bursts.

## Synchrocyclotron

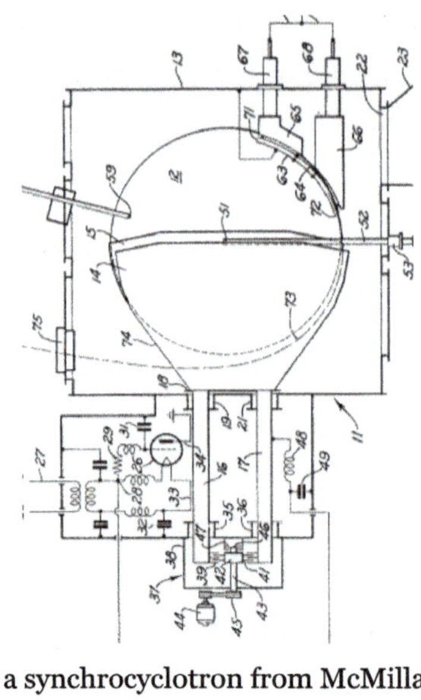

Sketch of a synchrocyclotron from McMillan's patent.

A synchrocyclotron is a special type of cyclotron, patented by Edwin McMillan, in which the frequency of the driving RF electric field is varied to compensate for relativistic effects as the particles' velocity begins to approach the speed of light. This is in contrast to the classical cyclotron, where this frequency is constant.

There are two major differences between the synchrocyclotron and the classical cyclotron. In the synchrocyclotron, only one *dee* retains its classical shape, while the other pole is open. Furthermore, the frequency of oscillating electric field in a synchrocyclotron is decreasing continuously instead of kept constant so as to maintain cyclotron resonance for relativistic velocities. One terminal of the oscillating electric potential varying periodically is applied to the dee and the other terminal is on ground potential. The protons or deuterons to be accelerated are made to move in circles of increasing radius. The acceleration of particles takes place as they enter or leave the dee. At the outer edge, the ion beam can be removed with the aid of electrostatic deflector. It was possible to produce 200 MeV deuterons and 400 MeV α-particles with the first synchrocyclotron.

## Differences From the Classical Cyclotron

In a classical cyclotron, the angular frequency of the electric field is given by

$$\omega = \frac{qB}{m},$$

Where $\omega$ is the angular frequency of the electric field, $q$ is the charge on the particle, $B$ is the magnetic field, and $m$ is the mass of the particle. This makes the assumption that the particle is classical, and doesn't experience relativistic phenomena such as length contraction. These effects start to become significant when $v$, the velocity of the particle greater than $\approx \frac{c}{3}$. To correct for this, the relativistic mass is used instead of the rest mass; thus, a factor of $\gamma$ multiplies the mass, such that

$$\omega = \frac{qB}{m\gamma},$$

where

$$\gamma = \frac{1}{\sqrt{1 - \frac{v^2}{c^2}}}.$$

This is then the angular frequency of the field applied to the particles as they are accelerated around the synchrocyclotron.

## Advantages

The chief advantage of the synchrocyclotron is that there is no need to restrict the number of revolutions executed by the ion before its exit. As such, the potential difference supplied between the dees can be much smaller.

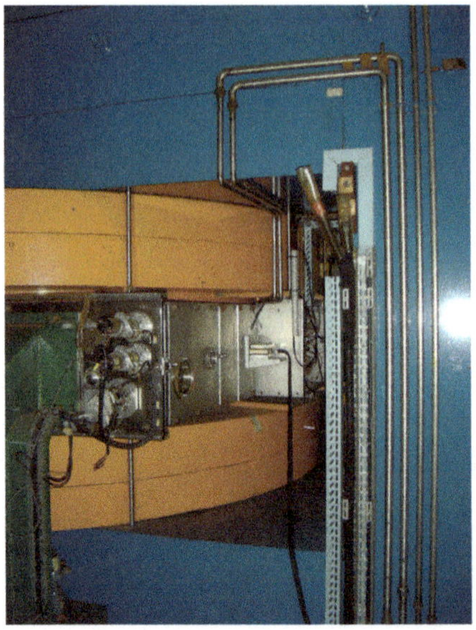

A part of the former Orsay synchrocyclotron

The smaller potential difference needed across the gap has the following uses:

1. There is no need for a narrow gap between the dees as in the case of conventional cyclotron, because strong electric fields for producing large acceleration are not required. Thus only one dee can be used instead of two, the other end of the oscillating voltage supply being connected to earth.

2. The magnetic pole pieces can be brought closer, thus making it possible to increase greatly the magnetic flux density.

3. The frequency valve oscillator is able to function with much greater efficiency.

## Disadvantages

The main drawback of this device is that, as a result of the variation in the frequency of the oscillating voltage supply, only a very small fraction of the ions leaving the source are captured in phase-stable orbits of maximum radius and energy with the result that the output beam current has a low duty cycle, and the average beam current is only a small fraction of the instantaneous beam current. Thus the machine produces high energy ions, though with comparatively low intensity.

The next development step of the cyclotron concept, the isochronous cyclotron, maintains a constant RF driving frequency and compensates for relativistic effects by increasing the magnetic field with radius. Isochronous cyclotrons are capable of producing much greater beam current than synchrocyclotrons. As a result, isochronous cyclotrons became more popular in the research field.

## Current Developments

Synchrocyclotrons are attractive for use in proton therapy because of the ability to make compact systems using high magnetic fields. Medical physics companies Ion Beam Applications and Mevion Medical Systems have developed superconducting synchrocyclotrons that can fit comfortably into hospitals.

## FFAG Accelerator

A Fixed-Field Alternating Gradient accelerator (FFAG) is a circular particle accelerator concept on which development was started in the early 50s, and that can be characterized by its time-independent magnetic fields (*fixed-field*, like in a cyclotron) and the use of strong focusing (*alternating gradient*, like in a synchrotron). Thus, FFAG accelerators combine the cyclotron's advantage of continuous, unpulsed operation, with the synchrotron's relatively inexpensive small magnet ring, of narrow bore.

Although the development of FFAGs had not been pursued for over a decade starting from 1967, it has regained interest since the mid-1980s for usage in neutron spallation sources, as a driver for muon colliders and to accelerate muons in a neutrino factory since the mid-1990s.

The revival in FFAG research has been particularly strong in Japan with the construction of several rings. This resurgence has been prompted in part by advances in RF cavities and in magnet design.

## History

## First Development Phase

The Michigan Mark I FFAG accelerator. This 400KeV electron accelerator was the first operational FFAG accelerator. The large rectangular part on the right is the betatron transformer core.

The idea of fixed-field alternating-gradient synchrotrons was developed independently in Japan by Tihiro Ohkawa, in the United States by Keith Symon, and in Russia by Andrei Kolomensky. The first prototype, built by Lawrence W. Jones and Kent M. Terwilliger at the University of Michigan used betatron acceleration and was operational in early 1956. That fall, the prototype was moved to the Midwestern Universities Research Association (MURA) lab at University of Wisconsin, where it was converted to a 500 keV electron synchrotron. Symon's patent, filed in early 1956, uses the terms "FFAG accelerator" and "FFAG synchrotron". Ohkawa worked with Symon and the MURA team for several years starting in 1955.

Donald Kerst, working with Symon, filed a patent for the spiral-sector FFAG accelerator at around the same time as Symon's Radial Sector patent. A very small spiral sector machine was built in 1957, and a 50 MeV radial sector machine was operated in 1961. This last machine was based on Ohkawa's patent, filed in 1957, for a symmetrical machine able to simultaneously accelerate identical particles in both clockwise and counterclockwise beams. This was one of the first colliding beam accelerators, although this feature was not used when it was put to practical use as the injector for the Tantalus storage ring at what would become the Synchrotron Radiation Center. The 50MeV machine was finally retired in the early 1970s.

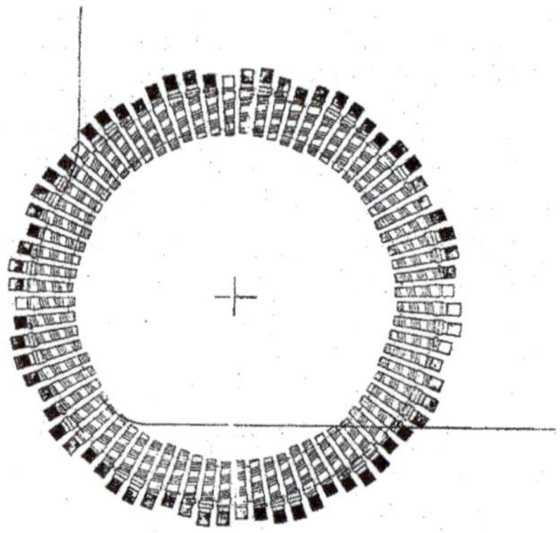

Layout of MURA FFAG

MURA designed 10 GeV and 12.5 GeV proton FFAGs that were not funded. Two scaled down designs, one for 720 MeV and one for a 500 MeV injector, were published.

With the shutdown of MURA which began 1963 and ended 1967, the FFAG concept was not in use on an existing accelerator design and thus was not actively discussed for some time.

## Continuing Development

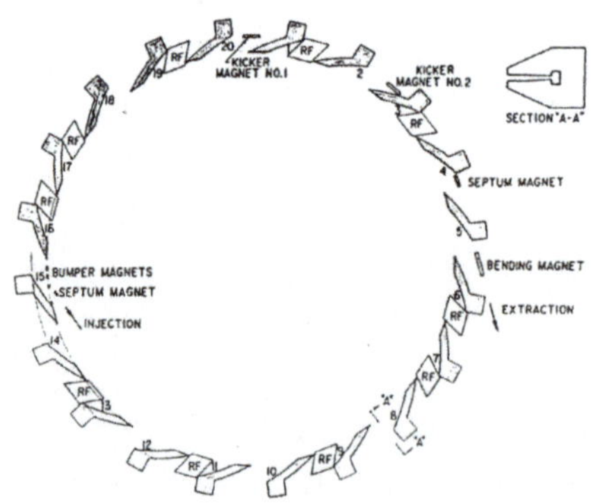


ASPUN ring (scaling FFAG). The first ANL design ASPUN was a spiral machine designed to increase momentum threefold with a modest spiral as compared with the MURA machines.

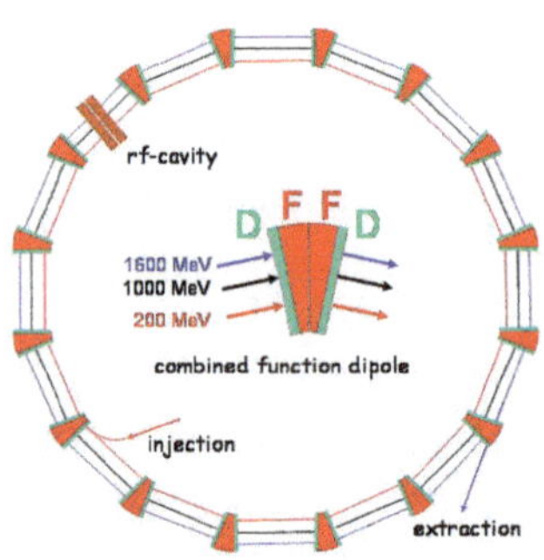


Example of a 16-cell superconducting FFAG. Energy: 1.6 GeV, average radius 26 m.

In the early 1980s, it was suggested by Phil Meads that an FFAG was suitable and advantageous as a proton accelerator for an intense spallation neutron source, starting off projects like the Argonne Tandem Linear Accelerator at Argonne National Laboratory and the Cooler Synchrotron at Jülich Research Centre.

Conferences exploring this possibility were held at Jülich Research Centre, starting from 1984. There have also been numerous annual workshops focusing on FFAG accelerators at CERN, KEK, BNL, TRIUMF, Fermilab, and the Reactor Research Institute at Kyoto University. In 1992, the European Particle Accelerator Conference at CERN was about FFAG accelerators.

The first proton FFAG was successfully construction in 2000, initiating a boom of FFAG activities in high-energy physics and medicine.

With superconducting magnets, the required length of the FFAG magnets scales roughly as the inverse square of the magnetic field. In 1994, a coil shape which provided the required field with no iron was derived. This magnet design was continued by S. Martin *et al.* from Jülich.

In 2010, after the workshop on FFAG accelerators in Kyoto, the construction of the Electron Machine with Many Applications was completed at Daresbury Laboratory, UK. This was the first non-scaling FFAG accelerator. Non-scaling FFAGs are often advantageous to scaling FFAGs because large and heavy magnets are avoided and the beam is much better controlled.

## Scaling vs Non-scaling Types

The magnetic fields needed for an FFAG are quite complex. The computation for the magnets used on the Michigan FFAG Mark Ib, a radial sector 500 keV machine from 1956, were done by Frank Cole at the University of Illinois on a mechanical calculator built by Friden. This was at the limit of what could be reasonably done without computers; the more complex magnet geometries of spiral sector and non-scaling FFAGs require sophisticated computer modeling.

The MURA machines were scaling FFAG synchrotrons meaning that orbits of any momentum are photographic enlargements of those of any other momentum. In such machines the betatron frequencies are constant, thus no resonances, that could lead to beam loss, are crossed. A machine is scaling if the median plane magnetic field satisfies

$$B_r = 0, \quad B_\theta = 0, \quad B_z = ar^k f(\psi),$$

where

- $\psi = N\,[\tan\zeta\ \ln(r/r_0) - \theta]$,
- $k$ is the field index,
- $N$ is the periodicity,
- $\zeta$ is the spiral angle (which equals zero for a radial machine),
- $r$ the average radius, and
- $f(\psi)$ is an arbitrary function that enables a stable orbit.

For $k >> 1$ an FFAG magnet is much smaller than that for a cyclotron of the same energy. The disadvantage is that these machines are highly nonlinear. These and other relationships are developed in the paper by Frank Cole.

The idea of building a non-scaling FFAG first occurred to Kent Terwilliger and Lawrence W. Jones in the late 1950s while thinking about how to increase the beam luminosity in the collision regions of the 2-way colliding beam FFAG they were working on. This idea had immediate applications in designing better focusing magnets for conventional accelerators, but was not applied to FFAG design until several decades later.

If acceleration is fast enough, the particles can pass through the betatron resonances before they have time to build up to a damaging amplitude. In that case the dipole field can be linear with radius, making the magnets smaller and simpler to construct. A proof-of-principle *linear, non-scaling* FFAG called (EMMA) (Electron Machine with Many Applications) has been successfully operated at Daresbury Laboratory, UK.

## Vertical FFAGs

Vertical Orbit Excursion FFAGs (VFFAGs) are a special type of FFAG arranged so that higher energy orbits occur above (or below) lower energy orbits, rather than radially outward. This is accomplished with skew-focusing fields that push particles with higher beam rigidity vertically into regions with a higher dipole field.

The major advantage offered by a VFFAG design over a FFAG design is that the path-length is held constant between particles with different energies and therefore relativistic particles travel isochronously. Isochronousity of the revolution period enables continuous beam operation, therefore offering the same advantage in power that isochronous cyclotrons have over synchrocyclotrons. Isochronous accelerators have no longitudinal beam focusing, but this is not a strong limitation in accelerators with rapid ramp rates typically used in FFAG designs.

The major disadvantages include the fact that VFFAGs requires unusual magnet designs and currently VFFAG designs have only been simulated rather than tested.

## Applications

FFAG accelerators have potential medical applications in proton therapy for cancer, as proton sources for high intensity neutron production, for non-invasive security inspections of closed cargo containers, for the rapid acceleration of muons to high energies before they have time to decay, and as "energy amplifiers", for Accelerator-Driven Sub-critical Reactors (ADSRs) / Sub-critical Reactors in which a neutron beam derived from a FFAG drives a slightly sub-critical fission reactor. Such ADSRs would be inherently safe, having no danger of accidental exponential runaway, and relatively little production of transuranium waste, with its long life and potential for nuclear weapons proliferation.

Because of their quasi-continuous beam and the resulting minimal acceleration intervals for high energies, FFAGs have also gained interest as possible parts of future muon collider facilities.

## Status

In the 1990s, researchers at the KEK particle physics laboratory near Tokyo began developing the FFAG concept, culminating in a 150 MeV machine in 2003. A non-scaling machine, dubbed PAMELA, to accelerate both protons and carbon nuclei for cancer therapy has been designed. Meanwhile, an ADSR operating at 100 MeV was demonstrated in Japan in March 2009 at the Kyoto University Critical Assembly (KUCA), achieving "sustainable nuclear reactions" with the critical assembly's control rods inserted into the reactor core to damp it below criticality.

## Generation of Energetic Particles in Accelerators

The science and technology of radiation detection has occupied a significant portion of modern science and technology because of its great impact in numerous applications in various fields such as information of material down to parts per million, study of astrophysical objects, archaeology and geology, biophysical imaging with high sensitivity.

There are three major criteria behind the design of a good nuclear detector, i.e. (i) choosing an appropriate material as detection medium primarily based on ion –mater interaction, (ii) conversion of incident energy in reliable electrical signal and (iii) design an appropriate electronic circuit to process this electrical signal with minimum noise.

It is also important to know how the energy of the particles is enhanced before they interact with the detection medium. Nature obviously provides energy to different particles by several processes. For example, photons generated in atomic transition have typical energies in the range of electron volt (eV), X-ray photons have energy in the range of kilo electron volt (keV) and gamma ray photons have energy in the range of mega electron volt (MeV). Particles coming from cosmic ray shower have a range of energies. It is also possible to use artificial means to enhance the energy of the particles. Therefore, a natural question arises:

## How to Increase Energy of the Particles?

Highly energetic particles can be generated by a man made machines, known as particle accelerators. Here we will discuss basic principles of some of these particles accelerators.

Historically, the credit of design and use the first modern accelerators goes to Cockroft and Walton, who in 1932 demonstrated the generation of 400 keV protons followed by the first ever artificial nuclear disintegration- that of lithium nucleus by means of these protons for the first time. The basic principle of operation of this accelerator was to charge a large number of condensers connected in parallel and discharge in series when the voltage across them becomes equal to the sum of the voltages of the individual condensers.

Almost at the same time, an electrostatic accelerator was designed by Van de Graff, popularly known as Van de Graff generator. In this generator, a charge Q is transported to a huge condenser C, resulting in a high voltage, $V = Q/C$, which is used to accelerate the ions generated in an ion source. The figure shows the basic features of this generator. A high voltage is used to spray positive charges on an insulating belt, which is driven by a motor to carry the charges continuously to a dome shaped terminal, where they are removed. Positive ions, like protons, deuterons, alpha particles etc. are produced in ion source and accelerated in the evacuated accelerating tube and gains energy qV, if it attains a charge state q. Eventually they are accelerated in the evacuated accelerating tube. Energy of the order of few tens of MeV can be generated in this machine.

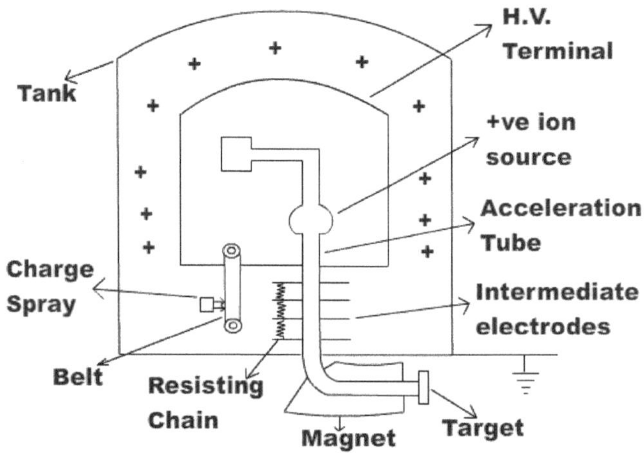

Schematic diagram of a Van de Graff generator

In order to enhance the energy a Tandem Van de Graff accelerator is designed, where two insulating columns are used one after another (in tandem) , in one tank so that the particles gain an energy twice that of Van de Graff generator. There are two types namely namely,

(i) folded type and

(ii) straight line type tandem used for acceleration purpose.

During same time, a concept of a linear resonant accelerator, popularly known as linear accelerator or linac was suggested and later on various heavy ion linacs were designed and constructed. The first successful linac was developed by Louis Alveraz and co workers, after World War II, which successfully energised protons upto 32 MeV. The basic principle of linac is that it accelerates particles in a straight line by means of an oscillating (radiofrequency) electric field, which provides a series of boosting kicks in correct phase at a series of electrode gaps.

The emergent beam from linac is highly collimated with high intensity.

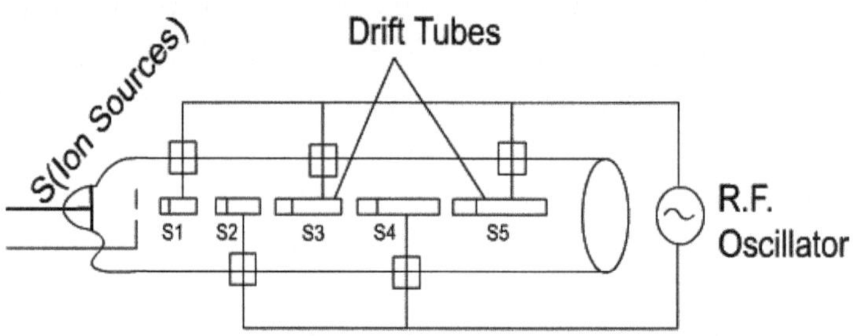

Schematic diagram of a Van de Graff generator

The ions get accelerated only in the gaps between two tubes where they are acted upon by the electric field present between the tubes , but travel with constant velocities in the field free space inside the drift tubes. Suppose the ion source S, some positive ions are generated in the ion source S . First these positive ions pass through the cylinder S1 to the gap 1 between S1 and S2. If q be the charge of each ion, the increment in their energy due to potential V is qV. Now again within S2 the ions move with constant speed and come to the next gap between S2 and S3 and get accelerated in the gap provided they arrive in the gap when S2 is positive w.r.to S3. In this way the acceleration process goes on with the alternate tubes connect electrically. Suppose the distance between the two gaps is l, and v be the velocity of the ion then,

$$l/v = T/2 = 1/2f$$

where T is the time period and f is the frequency of the oscillator. Length of the successive tubes keeps increasing to be in phase for acceleration. Length of tube and RF field is adjusted in such that ion ion exiting a tube sees 1800 phase change in RF field with respect to its point of entry.

Energy gained by the ions due to n accelerations in n gaps would be,

$$E = nqv = 1/2\ Mv^2$$

These expressions get modified due to relativistic correction* when the particle energy becomes quite high.

Both Van de Graff generator and linac are electrostatic accelerators.

Now we will discuss about an electromagnetic accelerator having circular geometry and popularly known as cyclotron. Cyclotron was first devised by Lawrence and Livingstone of University of California at Barkeley. This accelerator basically consists of two flat semi circular hollow metal boxes known as dees (D) (because of their shape) having diametric edges parallel and slightly separated from each other.

They are connected to a radio frequency oscillator, which essentially generates a high frequency ($10^6$ Hz) alternating potential between the dees, which act as electrodes. The dee structure is placed between the poles of a large electromagnet. An ion source is kept at the centre between the dees. Considering it generates protons , the mechanism of acceleration is explained based on the figure:

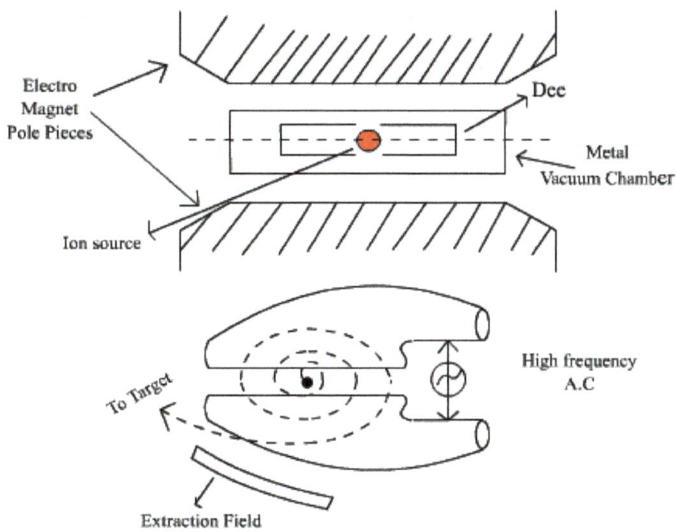


Schematic diagram of cyclotron

when D1 is negative, protons being positively charged are attracted towards it. Inside D1 the protons make a circular path say of radius R due to magnetic field. If the magnetic field strength is B, and the velocity of proton is v, then the radius R can be determined from the formula

$$m_p v^2/R = Bev$$

where $m_p$ and e are the mass and charge of a proton respectively. Inside the dee the protons are in a electric field free zone and therefore moves with a constant velocity. After travelling one complete semicircular path it comes at the edge of D1. If now the potential in the dees is adjusted in such a way that D2 becomes negative and D1 is positive , the proton will be attracted towards D2, accelerated and gaining further speed will enter inside D2. In a similar manner it will make a semicircular path inside D2 and come out at the edge of D2, where if the direction of field is changed will gain further energy and will enter insideD1. This process will continue and the proton will gain very high energy through several rotations. When it comes near the circumference of the dees, an electric field can be applied to extract the proton beam out and to focus it to the target. Now the maximum kinetic energy of the particle coming out of the machine can be calculated from the following formula:

$$(K.\ E.)_{max} = ½\, m_p v_m^2 = ½\, (R_m eB/m_p)^2 = e^2 B^2 R_m^2 / 2m_p$$

Cyclotron can energise charged particles to several MeVs, however, it suffers from a limitation due relativistic enhancement of mass which eventually start slowing down the particle.

I am not going to discuss this matter in detail. However, this problem can be overcome in a slightly modified design in another accelerator known as synchrocyclotron.

I have given here just a flavour about some basic accelerators, which are under routine use in different laboratories across the globe to generate accelerated particles. In India you will find such accelerators in the laboratories like Inter University Accelerator Centre (IUAC), New Delhi, Variable Energy Cyclotron Centre (VECC), Kolkata, Indira Gandhi Centre for Atomic Research (IGCAR), Kalapakkam etc. where basic researches on Nuclear Physics, Materials science, Atomic physics, Bio physics etc. are in progress. Apart from these accelerators there are some other accelerators, like, storage ring, particle collider, laser driven accelerators etc. where the energy of the particles can be increased upto GeV ($10^9$ eV) to TeV ($10^{12}$ 12 eV) range or higher are also available in different laboratories across the globe. The high energy nuclear physics or particle physics research is going on using these accelerators. You might have heard about large hadron collider (LHC), where experiments are in progress to get the signature of 'Higgs boson'.

## Interaction of Energetic Charged Particles with Matter

To understand the interaction of charged (alpha, beta, heavy ions etc.) and neutral particles (photons, neutron, neutrino) with matter is of paramount importance in nuclear detector technology. The energy deposited by these particles in the materials via various physical processes is converted to electrical signals, which eventually gives information about the particles.

## How Charge Particles Loose Energy in Matter?

When a charged particle interacts with any atoms of any material medium, it loses its energy via two mechanisms :

(i) by collision with the atom and

(ii) by exciting and ionizing the target atom.

The first process is an elastic collision process and popularly known as nuclear energy loss and the second is an inelastic process and is known as electronic energy loss. In order to understand the nuclear energy loss process, let me first set up basic equations related to a momentum transfer process when two objects collide.

Consider a two body collision process where a particle of mass m 1 and velocity v, energy $E_0$ (= ½ $m_1 v^2$) hits another body (at rest) of mass $m_2$. After collision the projectile scatters at an angle α and the target atom recoils with an angle θ as shown in the figure.

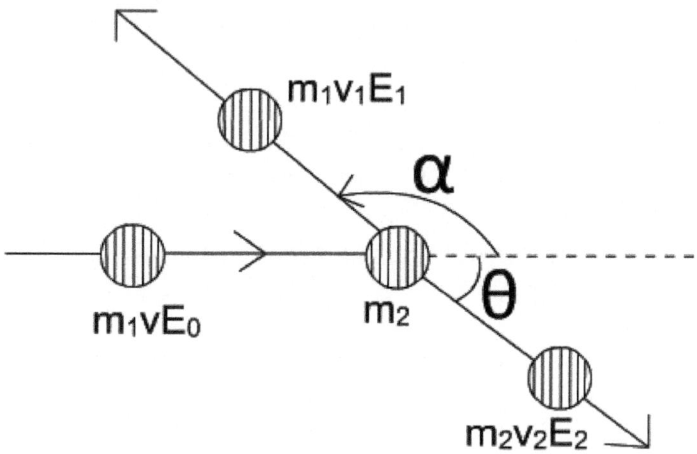


Schematic representation of elastic collision between two bodies of masses $m_1$ and m2.

The energy of the projectile becomes $E_1$ and that of target atom becomes $E_2$ after scattering and recoiling respectively. Let us now write down the energy-momentum conservation in the parallel and perpendicular to the direction of incidence as per the following mathematical equations:

Conservation of energy gives,

$$\frac{1}{2}m_1v^2 = \frac{1}{2}m_1v_1^2 + \frac{1}{2}m_2v_2^2$$

Conservation of momentum along incident ion direction gives,

$$m_1v = m_1v_1 \cos a + m_2v_2 \cos\theta$$

Conservation of momentum perpendicular to the direction of incident ion gives,

$$0 = m_1v_1 \sin a - m_2v_2 \sin\theta$$

Eliminating q and $v_2$ from the above equations one finds the following relations:

$$\frac{v_1}{v} = \frac{|\pm(m_2^2 - m_1^2 \sin^2 a)^{1/2} + m_1 \cos a|}{(m_2 + m_1)}$$

and hence

$$\frac{E_1}{E_0} = \left[ \frac{(m_2^2 - m_1^2 \sin^2 a)^{1/2} + m_1 \cos a \,|}{m_2 + m_1} \right]^2$$

The energy ratio, called the kinematic factor $K = E_1/E_0$, shows that the energy after scattering is determined only by the masses of the particle and target atom and the scattering angle. For direct backscattering of 180° , the energy ratio has its lowest value given by

$$\frac{E_1}{E_0} = \left( \frac{m_2 - m_1}{m_2 + m_1} \right)^2$$

And at 90° is given by

$$\frac{E_1}{E_0} = \frac{m_2 - m_1}{m_2 + m_1}$$

In collisions where m 1 = m 2, the incident particle is at rest after the collision with all the energy transferred to the target atom. For α =1800 , the energy E 2 transferred to the target atom has its maximum value given by

$$\frac{E_1}{E_0} = \frac{4m_2 - m_1}{(m_2 + m_1)^2}$$

with general relation given by

$$\frac{E_2}{E_0} = \frac{4m_1 m_2}{(m_2 + m_1)^2} \cos^2 \theta \,.$$

Therefore, it is clear from the above analysis that incident particle loses its energy while travelling inside the material by elastic collision process. The energy loss per unit length (dE/dx) by the elastic collision process is known as nuclear energy loss and is denoted by $S_n$.

Similarly, another component of energy loss, i.e. due to inelastic collision with electrons can be determined. Let us make a simple picture of this collision process, which was considered by Bohr 1913.

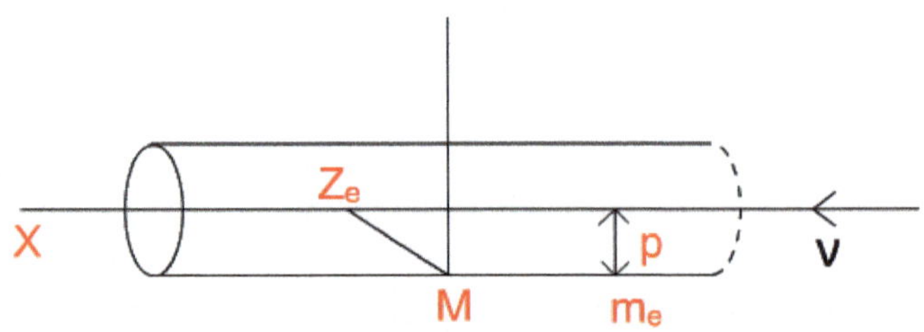


Schematic representation of collision of an ion with electrons of an atom

In his picture a particle of charge $Z_e$, mass M and velocity v passing through an atomic electron of mass m e at a distance p. As the particle passes through the material, it delivers its energy to the electrons of the atoms via Coulomb interaction and consequently loses its energy in the medium. As the excitation and ionization process enhances with increase in incident particle energy, the energy loss per unit path length $(dE/dx)_e$ or $S_e$ increases with increase in incident particle energy. However, at very high energy the probability of excitation and ionization drops down and hence the energy loss. Detailed mathematical formulation can be found in the literature. A simulation code based on Monte-Carlo simulation* process and considering the physical processes involved in both nuclear and electronic energy loss, have been developed by Z. F. Ziegler, which is known as transport of ions in matter (TRIM).

Let me now define the range and stragling of charge particles inside matter, two important physical quantities related to energy loss process. When charged particle passes through a material, due to energy loss by above two processes, it slows down and eventually stops after traversing certain distance inside the material. This distance is known as the range of that particle for that particular particle and material combination. It is obvious that range increases with increase in incident particle's energy. Now if we fix the energy and increase the density of the material, the range will decrease as the particle receives more atoms inside to interact with. However, as the irradiation is basically by many such charged particles, the energy loss is statistical process, i.e. some particles might stop before the average distance (range) it was supposed to go and some particles stop after this distance.

We can understand this point better from the following experiment:

Monoenergetic alpha particles from a collimated source are collected by a detector after passing through an absorber of variable thickness.

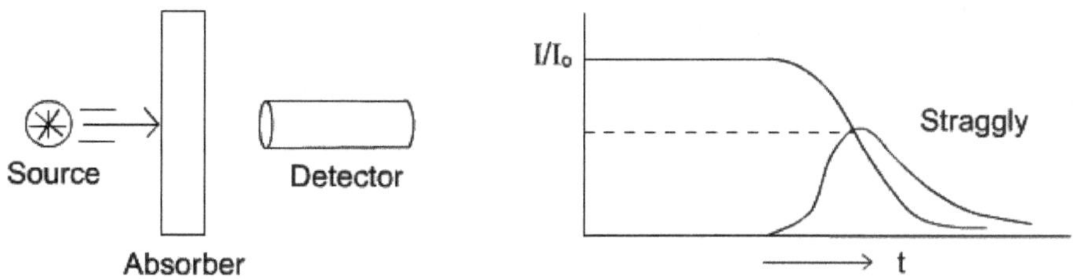


Schematic of alpha particle transmission experiment

For small values of absorber thickness, there is no change in the intensity of transmitted particle. With the increase in thickness, as the range is approached the intensity drops, though not as abruptly as expected the range to be a well-defined quantity. Instead the curve slopes down over a certain spread of thickness. This indicates that the energy loss process is a statistical phenomenon, as the two identical particles having same initial energy will not suffer the same number of collisions. This phenomenon is known as range straggling as indicated in the figure. This distribution can be considered Gaussian as a first approximation, and the mean value of the distribution is known as the mean range corresponds to the mid point as shown in the figure. Without considering multiple Coulomb scattering, the mean range of a particle of a given energy $T_0$, can be written by integrating the dE/dx formula,

$$S(T_0) = \int_0^{T_0} \left( \frac{dE}{dx} \right)^{-1} dE$$

This is an approximation of straight line range. If the multiple Coulomb scattering is considered, the path will be zigzag and will be slightly longer than the straight line path length.

## Interaction of Photons with Matter

The electromagnetic radiation corresponding to nuclear transition is gamma ray, which are highly energetic (typically mega electron volt, MeV) photons. They interact with materials by three different processes, namely, photoelectric effect, Compton scattering and pair creation. All these processes lead to partial or complete transfer of the photo energy to the electrons. These three processes eventually remove the photon from the beam entirely, either by absorption or by scattering. If a gamma ray/X-ray photon beam of intensity $I_0$ falls on a material, then it will be attenuated by these three processes as per the following relation:

$$I(d) = I_o \exp(-\mu d)$$

where d is the distance traverses by the photon beam inside the material and μ the absorption coefficient.

Let us now try to understand the above mentioned three physical processes by which the incident photons interact with the material and their contribution to the overall absorption cross section of photons inside the matter. The first process, i. e. photoelectric effect is basically is an absorption process where the incident photon interacts with atomic electron, gets absorbed, and subsequently the electron ejects out from one of the bound shells of the atom. If $E_{b.e.}$ is the binding energy of the electron in an atomic shell, then the kinetic energy of the ejected electron would be

$$K.E = hv.E_{b.e}$$

where hv is the energy of the incident photons. This is illustrated in the figure.

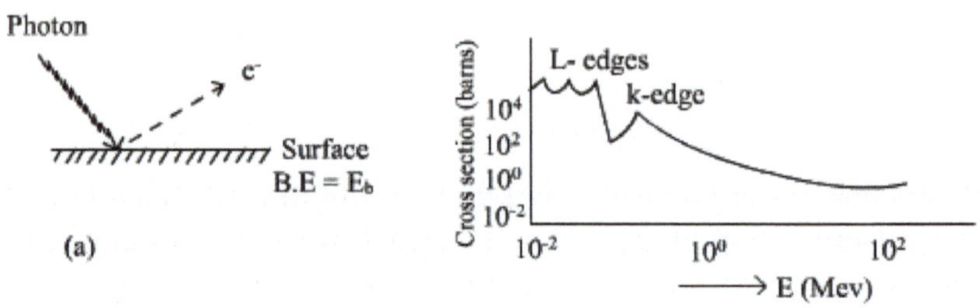


(a) Ejection of electrons from the surface due to photoelectric effect and
(b) variation of photoelectric effect cross section with incident ion energy.

For gamma rays of sufficient energy, the most probable origin of the photoelectron is the most tightly bound or K shell of the atom. The probability of occurrence of photoelectric effect or the cross section of photoelectric effect varies with incident photon energy as shown in the above. It is clear from this figure that the photoelectron emission probability increases with decrease in photon energy and attains a peak when it approaches to energy equal to the binding energy of K-shell of the atom. The cross section drops down just after that since K-electrons are not available for emission. Below this energy, the cross-section rises once again and dips as the L, M levels, etc. are passed. Photoelectric effect is dominant at relatively lower energy, and it is also enhanced for absorber material of high atomic number, Z.

Now we are going to discuss the second interaction process, i.e. Compton scattering, which is basically scattering of photons by free electrons. Though the electrons inside the material is bound, if the incident photon energy is higher enough with respect to the binding energy, this latter energy can be ignored and the electrons can be considered as essentially free. In the figure below, the kinetics of Compton scattering is shown, which clearly indicates that after scattering photon flies to a direction with an angle θ and the electron is kicked out from its position with an angle $\varphi$.

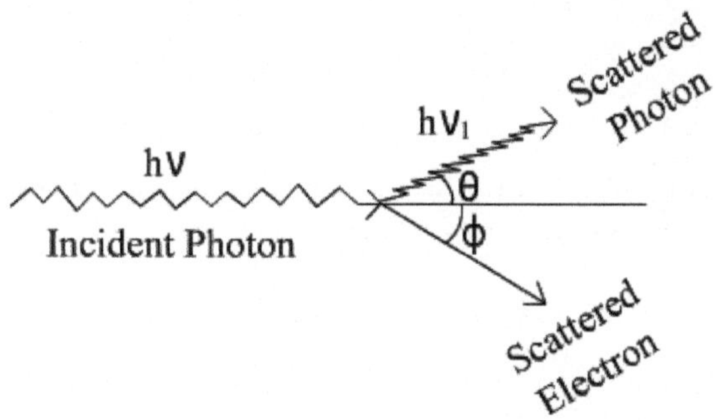


m2.8 Schematic representation of Compton scattering

If the incident photon energy is hν and scattered photon's energy is $h\nu_1$, then from energy-momentum conservation it can be shown that

$$h\nu_1 = h\nu\left[1 + h\nu\frac{(1-\cos\theta)}{m_0c^2}\right]$$

where $m_0c^2$ is the rest mass energy of electron , which is equal to 0.511 MeV. Klein-Nishima made a detailed theoretical work on the probability of Compton scattering and have shown that the probability varies linearly with Z value of the scatterer. As it is an angle dependent process, in a typical spectrum, it appears as a broad spectrum, known as Compton background.

The third process, known as pair creation is basically generation of electron-positron pairs when a gamma ray photon of energy 1.02 MeV or more passes through a material. In order to conserve momentum, this process occurs only near a third body, usually a nucleus. The minimum energy required for pair creation is 1.02 MeV, which is basically addition of rest mass energies of electron and positron. All the excess energy carried by photon above 1.02 MeV, required to create the pair, goes into kinetic energy shared by electron and positron. It has been shown1 that the probability of pair creation varies approximately square of the absorber's atomic number Z. The variation of pair production crosss section with incident photon energy is shown in the figure.

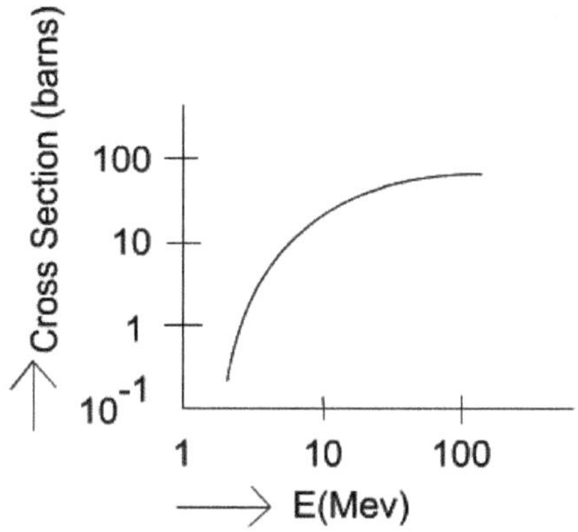


Pair production cross section with incident photon energy

Now when a gamma ray or a beam of gamma photons enters inside the material, it undergoes all these three processes simultaneously. Therefore, it is important to understand the relative contribution of individual process. This is depicted in the figure below, where cross section of these three processes versus incident photon energy is plotted. It is clear from this figure that photoelectric effect and Compton scattering are dominant at lower gamma energy, whereas at higher energy the contribution from pair production process enhances.

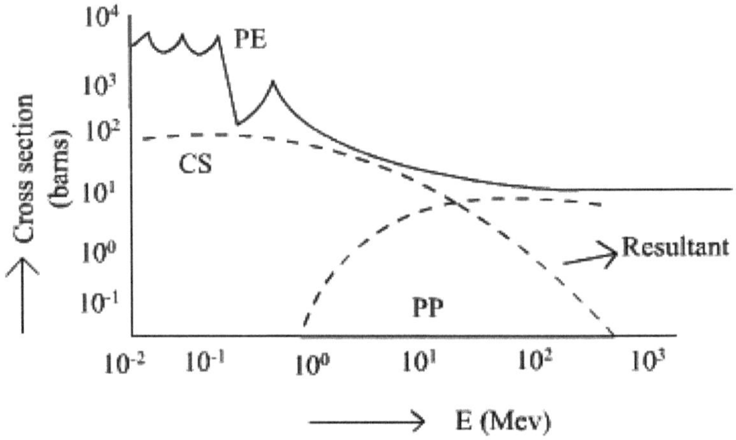


Cross section of photoelectric effect, Compton scattering and pair production versus incident gamma ray energy.

All the above three processes contribute additively to the total absorption coefficient of the gamma ray photons inside matter. Therefore, total cross-section

$$\sigma_{tot} = \sigma_{photo} + \sigma_c + \sigma_{pair}$$

Probability per unit length can be defined by multiplying tot and the density of atoms D and is given below:

$$\mu = \sigma_{tot} D = \sigma_{tot}\left(\frac{N_a \rho}{A}\right)$$

where $N_a$ is the Avogadro's number; ρ density of the material and A is the molecular weight. This is the total linear absorption coefficient and just the inverse of the mean free path of the photon.

Let us hold here for a moment and think about the situation when charged particles or photons interact with any compound or mixture. The question arises how to calculate energy loss by charged particles or absorption of photons in these cases?

In case of compounds, the energy loss can be obtained by averaging dE/dx over each element in the compound weighted by the fraction of each element (Bragg's Rule)[2]. Therefore,

$$\frac{1}{\rho}\frac{dE}{dx} = \frac{w_1}{\rho_1}\left(\frac{dE}{dx}\right)_1 + \frac{w_2}{\rho_2}\left(\frac{dE}{dx}\right)_2$$

Where $w_1$, $w_2$, etc. are the weight fractions of the elements 1,2 in the compound. If $n_i$ is the number of atoms of the $i^{th}$ element in the molecule m, then

$$w_i = \frac{n_i A_i}{A_n}$$

where $A_i$ is the atomic weight of the $i^{th}$ element and $A_n = \sum n_i A_i$.

Similarly, the Bragg rule for the absorption of photons can be written as:

$$\frac{\mu}{\rho} = \frac{\mu_1}{\rho_1} + \frac{\mu_2}{\rho_2}$$

## References


- Witman, Sarah. "Ten things you might not know about particle accelerators". Symmetry Magazine. Fermi National Accelerator Laboratory. Retrieved 21 April 2014
- Humphries, Stanley (1986). Principles of Charged Particle Acceleration. Wiley-Interscience. p. 4. ISBN 978-0471878780
- Nagai, Y.; Hatsukawa, Y. (2009). "Production of $^{99}$Mo for Nuclear Medicine by $^{100}$Mo(n,2n)$^{99}$Mo". Journal of the Physical Society of Japan. 78 (3): 033201. Bibcode:2009JPSJ...78c3201N. doi:10.1143/JPSJ.78.033201
- Cho, A. (June 2, 2006). "Aging Atom Smasher Runs All Out in Race for Most Coveted Particle". Science. 312 (5778): 1302. doi:10.1126/science.312.5778.1302. PMID 16741091
- "Atom smasher". American Heritage Science Dictionary. Houghton Mifflin Harcourt. 2005. p. 49. ISBN 978-0-618-45504-1
- Feder, T. (2010). "Accelerator school travels university circuit" (PDF). Physics Today. 63 (2): 20. Bibcode:2010PhT....63b..20F. doi:10.1063/1.3326981
- Abdelsalam, M.; Kustom, R. (July 1994). "Superconducting magnet design for Fixed-Field Alternating-Gradient (FFAG) Accelerator". 30 (4). IEEE. Retrieved 12 February 2017
- Hamm, Robert W.; Hamm, Marianne E. (2012). Industrial Accelerators and Their Applications. World Scientific. ISBN 978-981-4307-04-8
- Blewett, J. P. (1952). "Radial Focusing in the Linear Accelerator". Physical Review. 88 (5): 1197–1199. Bibcode:1952PhRv...88.1197B. doi:10.1103/PhysRev.88.1197
- Craddock, M.K. (2005). "New Concepts in FFAG Design for Secondary Beam Facilities and Other Applications" (PDF). Retrieved 12 February 2012
- Courant, E. D.; Livingston, M. S.; Snyder, H. S. (1952). "The Strong-Focusing Synchrotron—A New High Energy Accelerator". Physical Review. 88 (5): 1190–1196. Bibcode:1952PhRv...88.1190C. doi:10.1103/PhysRev.88.1190
- Ellis, J. LHC Safety Assessment Group; et al. (5 September 2008). "Review of the Safety of LHC Collisions" (PDF). Journal of Physics G. 35 (11): 115004. arXiv:0806.3414. Bibcode:2008JPhG...35k5004E. doi:10.1088/0954-3899/35/11/115004
- Wüstefeld, G. (14 May 1984). "2nd Jülich Seminar on Fixed Field Alternating Gradient Accelerators (FFAG)". Jülich. Retrieved 12 February 2017
- Widerøe, R. (17 December 1928). "Über Ein Neues Prinzip Zur Herstellung Hoher Spannungen". Archiv für Elektronik und Übertragungstechnik. 21 (4): 387. doi:10.1007/BF01656341
- "COSY - Fundamental research in the field of hadron, particle, and nuclear physics". Institute for Nuclear Physics. Retrieved 12 February 2017

- Chao, A. W.; Mess, K. H.; Tigner, M.; et al., eds. (2013). Handbook of Accelerator Physics and Engineering (2nd ed.). World Scientific. ISBN 978-981-4417-17-4
- Gahl and Flagg (2009).Solution Target Radioisotope Generator Technical Review. Subcritical Fission Mo99 Production. Retrieved 6 January 2013
- Peralta, E. A.; et al. "Demonstration of electron acceleration in a laser-driven dielectric microstructure". Retrieved 2014-05-01
- Briezman, B. N.; et al. "Self-Focused Particle Beam Drivers for Plasma Wakefield Accelerators" (PDF). Retrieved 2005-05-13
- Daniel Clery (4 January 2010). "The Next Big Beam?". Science. 327 (5962): 142–143. Bibcode:2010Sci...327..142C. doi:10.1126/science.327.5962.142
- "Two circulating beams bring first collisions in the LHC" (Press release). CERN Press Office. November 23, 2009. Retrieved 2009-11-23
- Clery, D. (2010). "The Next Big Beam?". Science. 327 (5962): 142–144. Bibcode:2010Sci...327..142C. doi:10.1126/science.327.5962.142
- Jones, L.; Mills, F.; Sessler, A.; Symon, K.; Young, D. (2010). Innovation was not enough: a history of the Midwestern Universities Research Association (MURA). World Scientific. ISBN 9789812832832
- D. Trbojevic, E. Keil, A. Sessler. "Non-Scaling Fixed Field Gradient Accelerator (FFAG) Design for the Proton and Carbon Therapy" (PDF). Retrieved 12 February 2017

# 4

# Types of Nuclear Detectors

Nuclear detectors use the energy loss of energetic materials to gather information about particles. A few types of nuclear detectors are gas detectors, solid state detectors, scintillation detectors, etc. The section strategically encompasses and incorporates the major types of nuclear detectors and their applications, providing a complete understanding.

## Gas Detectors

The challenge in nuclear detector design is how to obtain information about the incident particles from this energy loss and make it useful to understand various properties of the incident particle. In the simplest manner we can think that the energy given by the incident particles inside the medium produces an electron-positive ion pairs and those can be collected by electrodes eventually allowed to flow it through a circuit generating a current/voltage signal, one can obtain information about the energy of the incident particle. Let us first discuss the response of a gaseous medium under the application of an electric field. The response is basically the number of ion pairs collected in the electrodes placed in the gaseous medium. In the figure below, the number of ion pairs generated in a gas medium versus applied voltage for alpha and beta particles interacting with the medium are depicted.

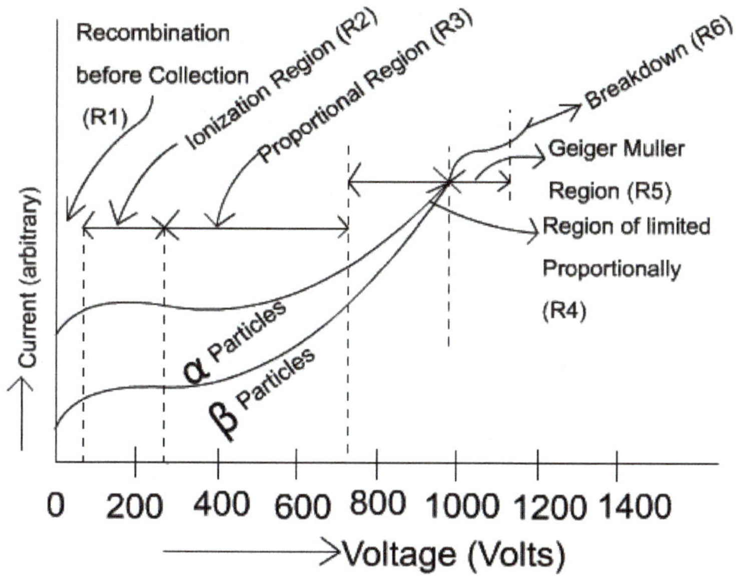


V-I characteristics of a gas medium in a single wire gas chamber geometry.

Initially at zero applied voltage, a very low current in the outer circuit is observed. This is because of the ion pairs generated due to the energy deposited by charged particles and collected by the electrodes. Therefore, to generate a proper signal it is important to apply voltage. As shown in the above figure there are six regions in this characteristics curve with increasing voltage. In the first region (R1) where applied voltage is very low the ion pairs get recombined before collected in the respective electrodes. With increase in applied voltage, the recombination force is overcome and the ion pairs are collected, thereby increasing current. After a certain voltage, all ion pairs are collected in the electrodes and saturation in current is observed. This region, designated as R2 is known as ionization region, where the ion pairs generated by primary ionization process are completely collected. Application of further voltage will result in an electric field capable enough to accelerate the freed electrons which will ionize further gas molecules in the chamber. This is known as secondary ionization, which eventually turns to ionization avalanche or cascade if the secondary electrons cause further ionization of gases. The number of electron ion pair in the avalanche is proportional to the number of primary ionization and generate a proportional amplification of the current, with a multiplication factor depending on the working voltage, V. This factor can be as high as $10^6$ so that the output signal is much larger than that from an ionization chamber. A detector working under this operating condition is known as proportional chamber and the region of operation is known as proportional region (R3) as shown in the above figure. In proportional chamber basically the positive electrode or anode is a wire inside a chamber and the entire chamber body acts as cathode. Now, the avalanche process occurs within a few radii of the anode wire, as the electric field is strongest near it. Further increase in voltage will create space charge, distorting the electric field near this region and hence the proportionality is no more strictly valid. This is known as region of limited proportionality (R4). Further increase in voltage creates very strong electric field resulting in discharge in the gas medium. Multiple avalanches spread throughout the anode length in which photons also contribute to ionization of gases. The output current thus gets completely saturated, always giving the same amplitude regardless of the energy of the initial event. Detectors working in this voltage region are known as Geiger-Müller counters (R5), which is characterized by a plateau over which the count rate varies little. A quenching gas is introduced in the medium in order to minimize the discharge by absorbing the photons. Finally, if the voltage is increased further a continuous breakdown occurs (R6) with or without radiation, which is of course avoided to prevent the damage of the detector.

Now before going to discuss on detectors, let us first know the general properties of radiation detectors.

(A) Detector Sensitivity

The main purpose of the nuclear detector is to produce a usable signal corresponding to certain radiation and energy. All detectors are not useful to all types of radiations and energy ranges. Sensitivity of the detector depends on the cross section of the ionizing reaction, detector mass, inherent noise etc., which basically defines the capability of the detectors to generate signals corresponding to certain radiation and energy. A certain minimum amount of ionization is required to generate a usable signal. This lower limit is determined by the noise from the detector and the electronics associated with it.

(B) Detector Response

In a detector, the amount of ionization is proportional to the energy it loses in the sensitive volume. If the radiation is completely absorbed, ionization gives a measure of energy of the radiation.

Output signal of the detector is generally in the form of a current pulse. The charge contained in this pulse basically measures the amount of ionization and hence the radiation energy. The relation between the radiation energy and the total charge or pulse height of the output signal is known as the response of the detector.

(C) Detector Resolution

If a mono energetic radiation is sent to a detector, one would expect a delta function kind of spectrum as output signal. However, in reality a distribution (Gaussian like) is seen with some finite width. This width occurs mainly due to statistical fluctuations in the number of ionization and excitation produced in the detection. Apart from this, intrinsic noise of the system and the errors of the instruments associated with the detector also cause the broadening. All these factors overall decide the resolution of the detection system i.e. the extent to which detectors can distinguish two close lying energies. For monoenergetic radiation, resolution is usually given in terms of the full width at half maximum (FWHM) of the peak. The FWHM is illustrated in the figure and is defined as the width of the distribution at a level that is just half the maximum ordinate of the peak intensity . Energies, which are closer than this interval, are usually considered un-resolvable. Resolution is defined as the ratio of FWHM and the centroid of the peak height Ho.

Therefore, Resolution,

$$R = \frac{FWHM}{H_0}$$

Equation m2.23 is usually expressed in percentage. A NaI detector has about 8% or 9% resolution for 1 MeV incident energy, for example, while Ge detectors have resolution of about 0.1%.

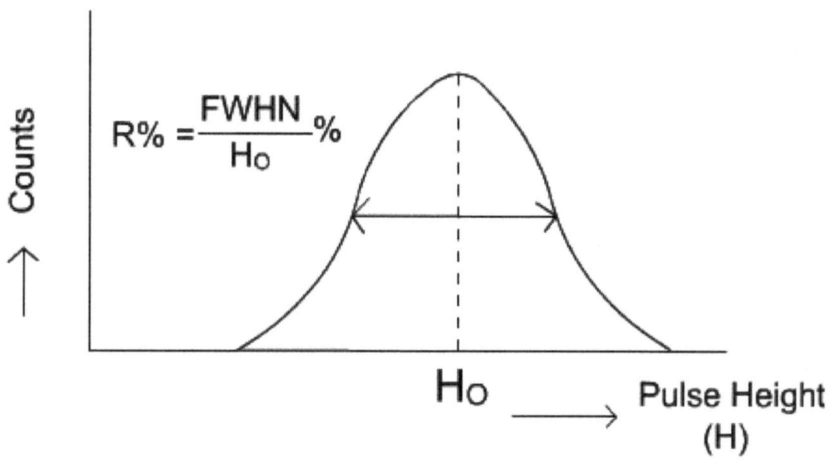


Resolution of a detector is defined in terms of FWHM and $H_0$.

(D) Detector Efficiency

A detection system has certain capacity to generate signal corresponding to the events emitted by the source. The absolute or total efficiency of a detector is defined as that fraction of events emitted by the source that is actually registered by the detector, i.e.

$$\xi_{tot} = \text{events registered/events emitted by the source}$$

It depends upon the detector geometry and the probability of an interaction in the detector. There is another efficiency of the detector known as intrinsic efficiency, which is defined as the ratio of events registered to the events impinging on the detector. Therefore,

$$\xi_{int} = \text{events registered/events impinging on the detector}$$

This probability depends on the interaction cross section of the incident radiation on the detector medium. Therefore, it is a function of the type of radiation, its energy and the detector material. Considering a cylindrical symmetry of the detector it can be shown that the geometrical efficiency depends on the geometrical configuration of the detector and source and also on the angular distribution of the incident radiation.

(E) Detector Dead time

For all detection system a minimum amount of time is required to generate an electrical signa corresponding to a radiation event and to separate signals corresponding to two different events. This time limit is decided either by the inner mechanism of the detector or by the associated electronics. This minimum time separation is usually called the dead time of the counting system. During this time, the detection system is insensitive and any event arriving during this period is lost. In case of high counting rates the dead time losses can become rather severe. For measurement of dead time in a typical detection system go through the references.

## Solid-state Nuclear Track Detector

A solid-state nuclear track detector or SSNTD (also known as an etched track detector or a dielectric track detector, DTD) is a sample of a solid material (photographic emulsion, crystal, glass or plastic) exposed to nuclear radiation (neutrons or charged particles, occasionally also gamma rays), etched, and examined microscopically. The tracks of nuclear particles are etched faster than the bulk material, and the size and shape of these tracks yield information about the mass, charge, energy and direction of motion of the particles. The main advantages over other radiation detectors are the detailed information available on individual particles, the persistence of the tracks allowing measurements to be made over long periods of time, and the simple, cheap and robust construction of the detector.

The basis of SSNTDs is that charged particles damage the detector within nanometers along the track in such a way that the track can be etched many times faster than the undamaged material. Etching, typically for several hours, enlarges the damage to conical pits of micrometer dimensions, that can be observed with a microscope. For a given type of particle, the length of the track gives the energy of the particle. The charge can be determined from the etch rate of the track compared to that of the bulk. If the particles enter the surface at normal incidence, the pits are circular; otherwise the ellipticity and orientation of the elliptical pit mouth indicate the direction of incidence.

SSNTDs are commonly used to study cosmic rays, long-lived radioactive elements, radon concentration in houses, and the age of geological samples.

A material commonly used in SSNTDs is polyallyl diglycol carbonate (PADC), also known as Tastrak, CR-39 and CR39. It is a clear, colorless, rigid plastic with the chemical formula $C_{12}H_{18}O_7$. Etching to expose radiation damage is typically performed using solutions of caustic alkalis such as sodium hydroxide, often at elevated temperatures for several hours.

## Solid State Detector

Among various solids, semiconductors (like Si, Ge) have been used in a significantly to design solid state detectors (SSDs). Recently, compound semiconductors with higher energy band gap are also used as SSDs for specific applications. Although semiconductor detectors are more costly in terms of availability, processing and design as compared to gas detectors, they are widely used for charged and gamma particle spectroscopy because of very good energy resolution. The energy required to produce one electron-positive ion pair is almost 10 times less than that of gas ionization detector. Recently, nuclear scientists are making detectors combing semiconductor and gas detectors with a very special design to determine energy and position of the particles.

In the rest part of this chpter, the basic physics behind semiconductors, making a semiconductor junction and designing a semiconductor detector is discussed.

Let us start from the basic discussion on intrinsic and extrinsic semiconductor and semiconductor junction.

## What is An Intrinsic Semiconductor?

Intrinsic semiconductor is basically a pure semiconductor, where valance band is completely filled and conduction band* is empty at absolute zero. At this condition, Fermi level # lies exactly in the middle of the band gap. Electrons can cross the band gap and reach to conduction band when temperature of the material is enhanced. In this case a hole is formed at the place from where electron has left and therefore, the number of electrons in the conduction band is exactly equal to the number of holes in the valance band. If n is the concentration of electrons in the conduction band and p is that of valance band, then from the equilibrium condition,

$$n_i = p_i$$

where subscript i stands for intrinsic. The electron and hole density at room temperature are $1.5 \times 10^{10}$ cm$^{-3}$ in Si and $2.4 \times 10^{13}$ cm$^{-3}$ in Ge respectively. Considering $\sim 10^{22}$ atoms/cm$^3$ in these materials one can see only 1 in $10^9$ Ge and 1 in $10^{12}$ Si atoms are ionized. Now let us ask a question 'what is the transport process of the carriers so generated'? Contrary to the metallic conductor, where negatively charged electrons are the only charge carriers, both negatively charged electrons and positively charged holes contribute to the conductivity of the intrinsic semiconductor

Under the application of an electric field both the electrons and holes undergo a net migration. The motion will be the combination of a random thermal velocity and a net drift velocity parallel to the direction of the applied field. Electrons move preferentially in an opposite direction to the electric field vector while hole moves in the same direction. The drift velocities of electrons and holes can be represented by

$$v_e = \mu_e E$$

$$v_h = \mu_h E$$

where, E is the magnitude of the applied electric field, $\mu_e$ and $\mu_h$ are the mobilities of the electrons and holes respectively. Increase in electric field causes increase in drift velocity, which eventually attains saturation value and becomes independent of electric field. Many semiconductor detectors are operated at electric field values sufficiently high to result in saturated drift velocity for the charge carriers. Physically, saturation occurs because of the proportional fraction of the kinetic energy acquired by the electrons and holes are drained by collision with the lattice atoms. Because the saturated velocities are of the order of 107 cm /sec, the time required collecting the carriers over typical dimensions of 0.1 cm or less will be around 10 ns. Semiconductor detectors are therefore be fastest-responding of all radiation detector types. The typical values of mobilities of electrons and holes in Si are 1350 $cm^2$ /Vs and 480 cm $^2$/ Vs, respectively. The mobilities, of course determine the current in a semiconductor. Since the current density J = ρv, where ρ is the charge density and v the velocity, J in a pure semiconductor is given by

$$J = en_i(\mu_e + \mu_h)E$$

Moreover, J = σE, where σ is the conductivity, given by the relation

$$\sigma = en_i(\mu_e + \mu_h)$$

This also gives the resistivity which is just inverse of σ. The following example shows the resistivity of intrinsic Si at room temperature.

Example m2.1

Find the resistivity of intrinsic Si at room temperature.

$$\rho = \frac{1}{en_i(\mu_e + \mu_k)} = 230,000\,\Omega - cm[taking\ e = 1.6\times10^{-19}C, n_i = 1.5\times10^{10} / cm_3, \mu_e = 1350cm^2 / V.s, \mu, = 480cm^2 / V.s$$

From intrinsic to extrinsic semiconductor:

Let us now go from intrinsic to extrinsic semiconductors, their fabrication and basic properties. In intrinsic semiconductor crystals, the number of holes equals the number of electrons in the conduction band. The number of either of these charge carriers can be preferentially increased by adding impurities in it. Suppose that we incorporate few arsenic atoms in a silicon crystal. Arsenic atoms have five electrons in their outermost shells, silicon atoms have four. These shells have the configurations $4s^2 4p^3$ and $3s^2 3p^2$ respectively. When an arsenic atom replaces a silicon atom in silicon crystal, four of its electrons participate in covalent bonds with nearest neighbors. The fifth electron requires little energy ~0.05 eV to be detached and move around inside the crystal. As shown in the figure, arsenic as an impurity in silicon provides energy levels just below the conduction band. Such levels are known as donor levels, and the substance is called n-type semiconductor because electric current in it is carried by negative charges. The Fermi level also rises from the mid position of the energy gap due to the presence of the donor levels.

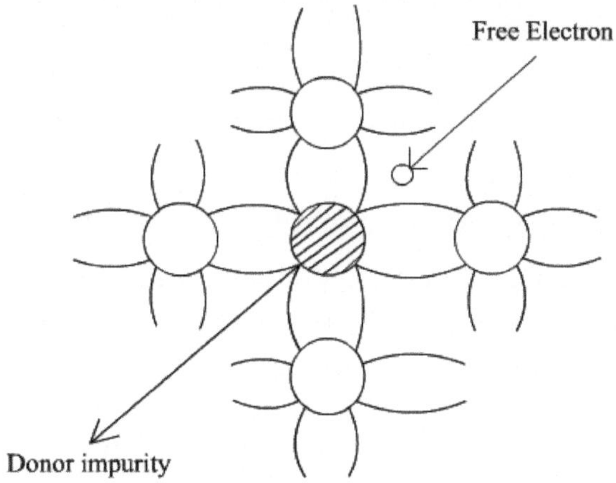


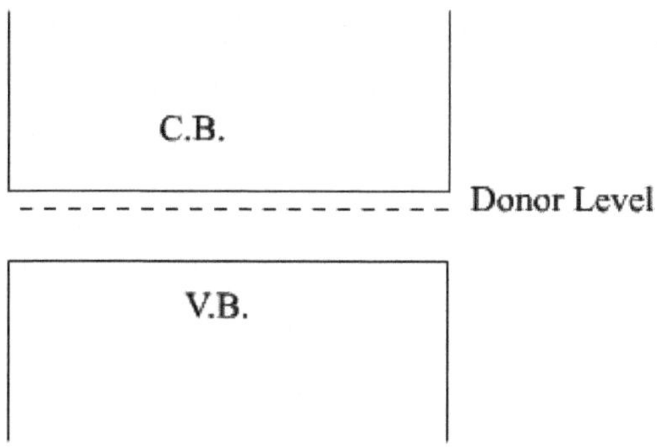


Doping of As in a silicon crystal provides donor levels in the normally forbidden band, producing an n-type semiconductor.

Now, if a trivalent atom gallium (Ga) is incorporated in a silicon crystal different characteristic occurs. The configuration of outer most electron shell of Ga is $4s^2 4p$, and their presence leaves vacancies called holes in the electron structure of the crystal. We see that Ga as an impurity in silicon provides energy levels called acceptor levels, just above the valance band and the Fermi energy lies below the middle of the energy gap. This is known as p-type semiconductor.

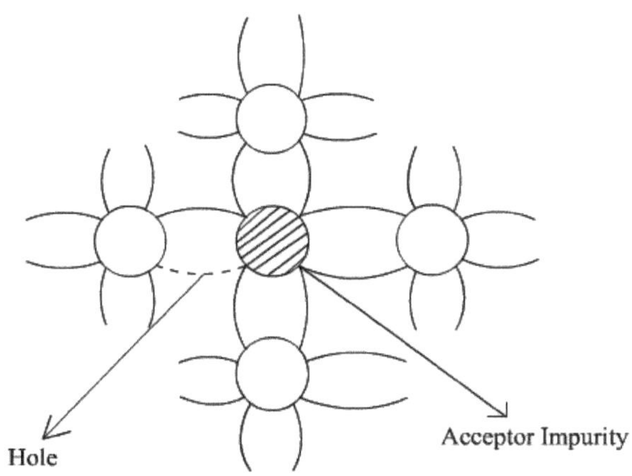

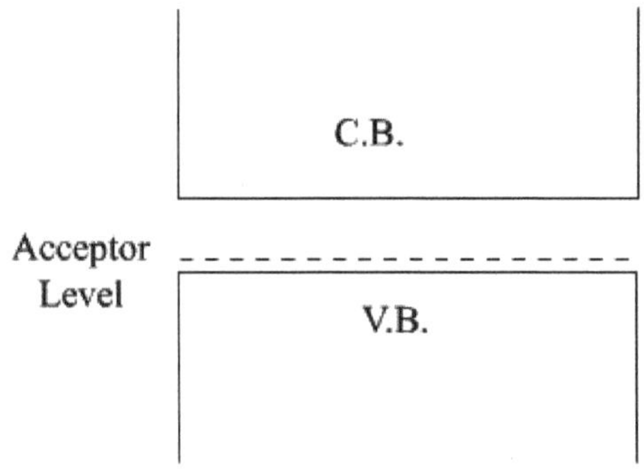


Doping of Ga in a silicon crystal provides acceptor levels in the normally forbidden band, producing a p-type semiconductor.

The amount of dopant used is generally very small with typical concentration being on the order of $10^{13}$ atoms/cm$^3$ . Since the atomic concentration of Si and Ge is of the order of $10^{22}$ atoms/cm$^3$ , this implies impurity is of only a few parts per billion. Regardless of the type of dopant, the concentration of electrons and holes obey a simple law of mass action at thermal equilibrium. If n is the concentration of electrons and p is the concentration of holes, then their product is

$$np = n_i^2 = AT^3 \exp\left(\frac{-E_g}{kT}\right)$$

where $n_i$ is the intrinsic concentration. Since the semiconductor is neutral, the negative and positive densities must be equal, so that

$$N_D + p = N_A + n$$

where $N_D$ and $N_A$ are donor and acceptor concentrations. In an n-type material, where N A =0 and n>>p, the electron density is therefore

$$n \cong N_D$$

i.e. the electron concentration is approximately the same as dopant concentration. Then the minority carrier concentration is

$$p = \frac{n_i^2}{N_D}$$

Therefore, the conductivity and resistivity of an n-type material becomes

$$\frac{1}{\rho} = \sigma \approx\sim eN_D\mu_e$$

An analogous result is found for p-type material.

From Extrinsic Semiconductor to Semiconductor Junction:

Now let us see how these extrinsic semiconductors are used to make a semiconductor junction and use them in real device like semiconductor detector. The function of semiconductor detectors depends on the formation of good semiconductor junction. For this purpose, one method for example is to diffuse sufficient p-type impurities into one end of a homogeneous bar of n type material so as to change that end into a p-type semiconductor. When a pn-junction is formed there is an initial diffusion of holes towards the n-region and similar diffusion of electrons into p-region due to difference in the concentration. The diffusing electrons fill up holes in the p-region while the diffusing holes capture electrons on the n-side, which eventually result in charge build-up to occur on either side of the junction. Since the p-region is injected with extra electrons it thus becomes negative while the n-region becomes positive. This creates an electric field gradient across the junction, which halts the diffusion process leaving a region of immobile space charge as shown schematically in figure (a). The corresponding electronic energy level is shown in the figure (b).

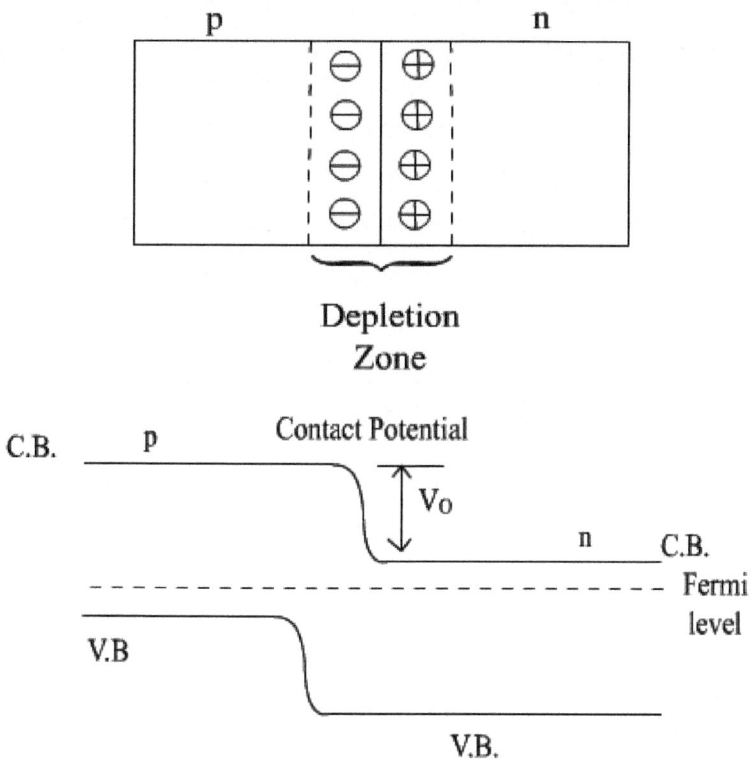


(a) Schematic diagram of an np junction, (b) diagram of electronic energy levels showing creation of a contact potential $V_0$

The region of changing potential is known as depletion zone or space charge region devoid of all mobile charge carriers. Any electron-hole created or entering in this zone will be swept out by the electric field. The width of the depletion zone, which depends on the concentration of n and p impurities, is very much important for detection purpose. The width (d) of the depletion zone can be calculated from simple electrostatic equations and can be shown to be equal to[1]

$$d = \left[ \frac{2\varepsilon V_0 (N_A + N_D)}{e N_A N_D} \right]$$

Where $N_A$ and $N_D$ are concentration of acceptor and donor impurities, e is the electronic charge and e is the dielectric constant of the medium. Because of its electrical configuration, the depletion layer also has certain capacitance, where A is the area of the depletion zone and d is the width.

$$C = \varepsilon \frac{A}{d}$$

For an unbiased junction the width of depletion region is very small and therefore stopping of high-energy particles is not possible. Due to small width of this zone the capacitance C is also quite high and hence increases the noise of the detector. Therefore, a reverse bias i.e. a negative voltage to the p-side and positive voltage to the n-side has to be applied in order to increase the depletion depth. This is illustrated in the figure.

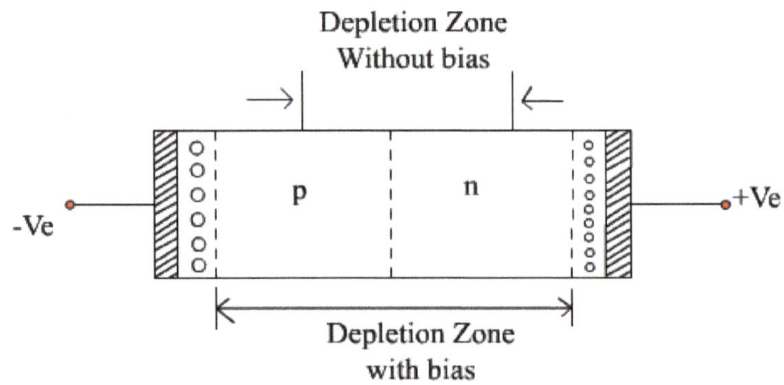


Semiconductor junction with and without reverse bias indicating broadening of depletion zone.

This voltage attracts the holes in the p-region away from the junction and similarly for the electrons in the n-region resulting an enlargement of the depletion zone. The Higher the external voltage, the wider the depletion zone. The higher external voltage will also provide efficient charge collection, which is otherwise very small for only intrinsic field. A simple calculation based on above equations shows that for n-type silicon a depletion layer greater than 1 mm can be obtained under a reverse bias of 300 V. This is a great improvement over a few tens of microns in an unbiased junction.

In order to collect charges produced by radiation, electrodes must be attached to the two sides of the junction. To make a direct ohmic metallic contact to the semiconductor is practically difficult as a junction extending to the semiconductor is formed. To prevent this formation, a heavily doped (typically ~$10^{20}$ atoms/$cm^3$ ) layer of n+ or p+ materials is used between the semiconductor and the metal leads.

The average energy (w) needed to create electron-hole pair is much smaller as compared to gas ionization detector. Like gases, the average energy at a given temperature is found to be independent of the type and energy of the radiation and depends only on the type of material. Due to low values of average energy, the number of charge carriers will therefore be almost an order of magnitude greater in these materials than in gases, and hence semiconductor detectors provide much higher resolution (i.e. lower FWHM in energy peak) . The typical values of average energies for Si and Ge are shown in the table.

Average energy for electron-hole creation in Si and Ge .

| Average energy (eV) | For Si | For Ge |
|---|---|---|
| 300 K | 3.62 | -------- |
| 77 K | 3.81 | .96 |

If the incident radiation of energy E is totally stopped in the depletion region, then E/w no. of electron-hole pair should be created. Considering the collection efficiency to be ε, a charge ε E/w will be collected on the electrodes. The observed voltage V on the electrode is then

$$V = \frac{Q}{C} = \frac{\varepsilon E}{wC}$$

where, C is the capacitance of the depletion region as shown earlier. It is clear from the above relation that the signal varies linearly with E. Moreover, since w is independent of particle type, the response is also independent of type of radiation. Now, if the depletion zone is smaller than the range of the radiation, it is clear then that a nonlinear response should be expected since the full energy is not totally deposited in the sensitive volume. In this case, instead of total energy, the energy loss DE is measured, which is a nonlinear function of E. Generally, semiconductor detectors are highly efficient (intrinsic efficiency ~100 %) for charged particle as only very few particles will fail to create ionization in the sensitive volume. The sensitivity is of course limited by the noise generated from leakage current in the detector and from the associated electronic circuitry. A second source is thermally generated electron-hole pairs originating from recombination and trapping centers in the depletion layer.

Semiconductor Junction as Radiation Detector:

Now we will see how the semiconductor junction is used as radiation detector. Semiconductor detectors can be fabricated in different process and configurations such as diffused junction detectors, surface barrier detectors, ion-implanted diode detectors, fully depleted detectors, passivated planar detectors etc. Here we shall discuss only about solid state surface barrier detectors (SSBD), the most widely used for charged particle detection. The basic design of this type of detector is the formation of semiconductor-metal junction, usually n-type semiconductor with gold and p-type semiconductor with aluminium. The junction so formed is known as Schottky barriers, where the depletion zone extending entirely into the semiconductor side. The depletion depth in a surface barriers detector can be calculated using the equation and for high resistivity silicon, depths of ~5 mm can be achieved. Generally SSB detectors are made at room temperature by first etching the silicon surface and then depositing a thin layer (~ 40 mg/cm2 ) of gold by evaporation. In this process, it is also necessary to allow the surface to oxidize slightly before deposition. The junction is then mounted in an insulating ring with metallized surfaces for electrical contact.

Surface barrier detectors are generally of two types - fully depleted and partially depleted. For fully depleted detectors, a gain in the collection time of charges can be obtained resulting in a faster signal risetime. In case of partially depleted detectors, an increase in bias also extends the depletion zone and thus the distance over which charges must be collected. SSBD's are very sensitive to surface contamination and care must be taken to keep the surface clean. If the detector is used in the vacuum chamber, attention must be paid to the possibility of presence of oil from the vacuum pump being deposited on the surface. There are suitable devices such as foreline trap, LN2 trap, oil mist eliminator etc. to get rid of these problems.

## Scintillation Detectors

We have seen two different detection media, i.e. gas and semiconductor. Although these two are two different media, the detection mechanism in these two has lot of similarities. When energetic particles enters inside in these media , electron-positive ion pairs (in gas detectors) or electron-hole pairs (in semiconductor detectors) are generated, which eventually creates a signal through a proper arrangement of electrodes.

## General Characteristics of Scintillation Detectors

A scintillation detection system consists of a thin window followed by the scintillator, which is optically coupled to a photomultiplier tube (PMT) either directly or via a light guide housed in a same assembly. The PMT converts the light falling on it into photoelectrons, which eventually gets further dynodes as shown in the figure. Finally the electrons are collected at anode form a signal using a suitable electronic circuit , giving rise to a current. The entire structure is properly shielded to protect from any unwanted radiation coming from any other source and also light loss through the boundaries. The design of the detection system is based on two important points - light collection and transport. It is important to collect as many of the emitted photons as possible and to efficiently couple them to the PMT. The main problem encountered in this detector is the loss due to transmission through the scintillator boundaries. Light emitted at any given point of the scintillator travels in all directions and only a fraction of it reaches to photocathode of PMT. Rest of it travels towards the scintillator boundaries from where a fraction of it reflects back and other part gets transmitted depending on the angle of incident. This is schematically shown in the figure. This loss reduces the efficiency and energy resolution of the detector.

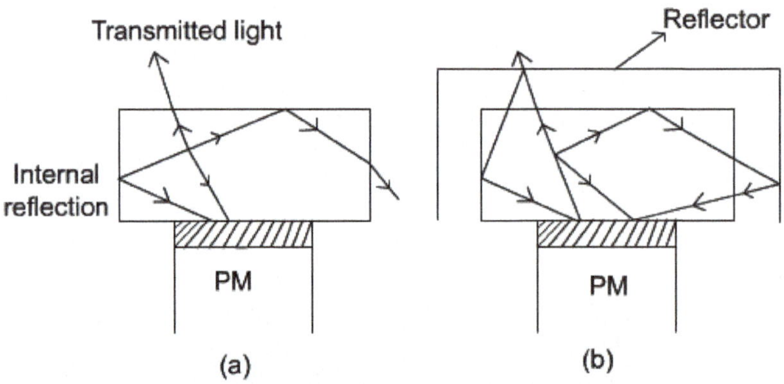


Light transmission in a typical scintillator

In order to improve light collection the boundary materials of high reflectivity are chosen. The most common materials used for this purpose are MgO, $TiO_2$, Al etc. These are usually form of a powder or as a white paint. It has been shown that Al and MgO retain relatively high reflectivity down to low wavelengths, $TiO_2$ drops off sharply at ~400 nm where it becomes a poor reflector. Apart from this, the maximization of internal reflection is another important task. It is known that light impinging at an angle greater than Brewester angle, $\theta_B$ where

$$\theta_B = \sin^{-1}\left(\frac{n_{med}}{n_{scint}}\right)$$

with $n_{med}$ being the refractive index of the surrounding medium and n $_{scin}$t that of the scintillator, total internal reflection occurs. The medium surrounding the scintillator should therefore have low refractive index value to minimize q B. For this reason, a layer of air should be maintained between the reflector and the scintillator. Next stage is the coupling between the scintillator and the PM, which has to be made in such a way so that maximum light transmits. The optical contact between scintillator and PM window is made through such a material whose index of refraction is as close as possible to both. In case of organic scintillator, silicon grease or oil is chosen, however, for inorganic material finding an appropriate material is difficult. Apart from this direct coupling, in many cases coupling is done through a light guide (or light pipe). Light guide works in the principle of internal reflection, that is the light entering from one end is 'guided' along the pipe by internal reflections back and forth between the interior wall.

The next stage of the scintillation detection system is photomultiplier tube (PMT), an electron tube device that converts light into a measurable electric current. The basic components of a PMT are schematically represented in the figure. It consists of a cathode made of photosensitive material, followed by an electron collection system, an electron multiplier section, and finally an anode from where the final signal is to be taken. All parts are usually housed in an evacuated glass tube. The phtocathode converts incident light into a current of electrons by the photoelectric effect. Efficiency of photoelectric conversion varies strongly with the frequency of the incident light and the structure of the material. The overall spectral response is expressed by the quantum efficiency, which is defined as the ratio of the umber of photoelectrons released to the number of incident photons on cathode. An equivalent quantity is the radiant cathode sensitivity, which is defined as

$$S(\lambda) = \frac{I_k}{P(\lambda)}$$

where $I_k$ is the photoelectric emission current from the cathode and $P(\lambda)$ is the incident radiant power. After emission from the photocathode, electrons in the PM are focused to the electron-multiplier system by an electron-optical input system. This focusing is accomplished through the application of an electric field in a suitable configuration.

The design of the electron-optical input system is based on the requirements of efficient collection of electrons i.e. as many electrons as possible must reach the electron-multiplier section regardless of the point of origin on the cathode. The electron-multiplier system amplifies the weak photocurrent by using a series of secondary electrodes or dynodes. The dynodes work in the principle of secondary electron emission, which is like photoelectric effect where photon is replaced by electron. The materials used for such purpose should have high secondary emission factor i.e. the average number of secondary electrons emitted per primary electron. Materials in common use today are Ag-Mg, Cu-Be, and Cs-Sb. Photomultiplier's generally have a wide range of possible working voltages extending over a 1000 V or more for an optimized gain and pulse height of the output signal.

## Nuclear Electronics

Nuclear electronics is a subfield of electronics concerned with the design and use of high-speed electronic systems for nuclear physics and elementary particle physics research, and for industrial and medical use.

Essential elements of such systems include fast detectors for charged particles, discriminators for separating them by energy, counters for counting the pulses produced by individual particles, fast logic circuits (including coincidence and veto gates), for identification of particular types of complex particle events, and pulse height analyzers (PHAs) for sorting and counting gamma rays or particle interactions by energy, for spectral analysis.

## Elementary Components

Some of the essential components that make up the elements of a nuclear electronic analysis system include:

- Detectors
- Preamplifiers
- Discriminators
- Coincidence and veto logic gates.
- Counters
- Pulse height analyzers

These elements were originally developed and built in the laboratories of the scientists doing the pioneering work in the field, but are nowadays designed, developed, and manufactured by a variety of specialized vendors:

- Cremat Inc.
- Apantec LLC.
- Berkeley Nucleonics
- Wiener Plein & Baus GmbH
- Canberra Industries now a division of Areva
- EG&G Ortec
- Oxford Instruments
- Stanford Research Systems
- Tennelec

5

# Nuclear Reactions and Reactors

Nuclear reaction is a reaction in which two or more nuclei or their components collide to form a nuclide, which attributes different characteristics. Also it is necessary to know that there is output of energy and not reverse effect. For this, the energy gain or loss is calculated which is called Q-value. The more the Q-value, the faster the reaction takes place. The diverse applications of nuclear reactions and reactors in the current scenario have been thoroughly discussed in this chapter.

## Nuclear Reaction

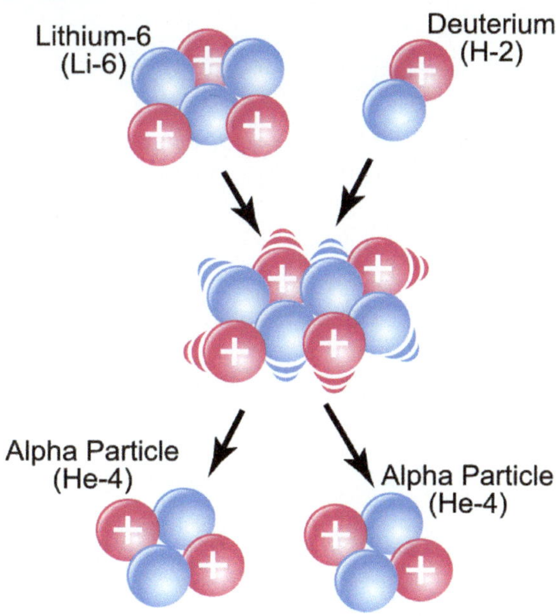


Lithium-6 – Deuterium Reaction

In this symbolic representing of a nuclear reaction, lithium-6 ($^6_3$Li) and deuterium ($^2_1$H) react to form the highly excited intermediate nucleus $^8_4$Be which then decays immediately into two alpha particles of helium-4 $^4_2$He. Protons are symbolically represented by red spheres, and neutrons by blue spheres.

In nuclear physics and nuclear chemistry, a nuclear reaction is semantically considered to be the process in which two nuclei, or else a nucleus of an atom and a subatomic particle (such as a proton, neutron, or high energyelectron) from outside the atom, collide to produce one or more nuclides that are different from the nuclide(s) that began the process. Thus, a nuclear reaction must cause a transformation of at least one nuclide to another. If a nucleus interacts with another nucleus or particle and they then separate without changing the nature of any nuclide, the process is simply referred to as a type of nuclear scattering, rather than a nuclear reaction.

In principle, a reaction can involve more than two particlescolliding, but because the probability of three or more nuclei to meet at the same time at the same place is much less than for two nuclei, such an event is exceptionally rare. "Nuclear reaction" is a term implying an induced change in a nuclide, and thus it does not apply to any type of radioactive decay (which by definition is a spontaneous process).

Natural nuclear reactions occur in the interaction between cosmic rays and matter, and nuclear reactions can be employed artificially to obtain nuclear energy, at an adjustable rate, on demand. Perhaps the most notable nuclear reactions are the nuclear chain reactions in fissionable materials that produce induced nuclear fission, and the various nuclear fusion reactions of light elements that power the energy production of the Sun and stars. Both of these types of reactions.

Nuclear reactions may be shown in a form similar to chemical equations, for which invariant mass must balance for each side of the equation, and in which transformations of particles must follow certain conservation laws, such as conservation of charge and baryon number (total atomic mass number). An example of this notation follows:

$$^{6}_{3}\mathrm{Li} + ^{2}_{1}\mathrm{H} \rightarrow ^{4}_{2}\mathrm{He} + ?.$$

To balance the equation above for mass, charge and mass number, the second nucleus to the right must have atomic number 2 and mass number 4; it is therefore also helium-4. The complete equation therefore reads:

$$^{6}_{3}\mathrm{Li} + ^{2}_{1}\mathrm{H} \rightarrow ^{4}_{2}\mathrm{He} + ^{4}_{2}\mathrm{He}.$$

or more simply:

$$^{6}_{3}\mathrm{Li} + ^{2}_{1}\mathrm{H} \rightarrow 2^{4}_{2}\mathrm{He}.$$

Instead of using the full equations in the style above, in many situations a compact notation is used to describe nuclear reactions. This style of the form A(b,c)D is equivalent to A + b producing c + D. Common light particles are often abbreviated in this shorthand, typically p for proton, n for neutron, d for deuteron, α representing an alpha particle or helium-4, β for beta particle or electron, γ for gamma photon, etc. The reaction above would be written as Li-6(d,α)α.

## History

In 1919, Ernest Rutherford was able to accomplish transmutation of nitrogen into oxygen at the University of Manchester, using alpha particles directed at nitrogen $^{14}\mathrm{N} + \alpha \rightarrow ^{17}\mathrm{O} + \mathrm{p}$. This was the first observation of an induced nuclear reaction, that is, a reaction in which particles from one decay are used to transform another atomic nucleus. Eventually, in 1932 at Cambridge University, a fully artificial nuclear reaction and nuclear transmutation was achieved by Rutherford's colleagues John Cockcroft and Ernest Walton, who used artificially accelerated protons against lithium-7, to split the nucleus into two alpha particles. The feat was popularly known as "splitting the atom", although it was not the modern nuclear fission reaction later discovered in heavy elements, in 1938 by the German scientists Otto Hahn and Fritz Straßmann.

## Energy Conservation

Kinetic energy may be released during the course of a reaction (exothermic reaction) or kinetic energy may have to be supplied for the reaction to take place (endothermic reaction). This can be calculated by reference to a table of very accurate particle rest masses, as follows: according to the reference tables, the ${}^{6}_{3}Li$ nucleus has a relative atomic mass of 6.015 atomic mass units (abbreviated u), the deuterium has 2.014 u, and the helium-4 nucleus has 4.0026 u. Thus:

- total rest mass on left side = 6.015 + 2.014 = 8.029 u;
- total rest mass on right side = 2 × 4.0026 = 8.0052 u;
- missing rest mass = 8.029 – 8.0052 = 0.0238 atomic mass units.

In a nuclear reaction, the total (relativistic) energy is conserved. The "missing" rest mass must therefore reappear as kinetic energy released in the reaction; its source is the nuclear binding energy. Using Einstein's mass-energy equivalence formula $E = mc^2$, the amount of energy released can be determined. We first need the energy equivalent of one atomic mass unit:

$$1\,\mathrm{u}\,c^2 = (1.66054 \times 10^{-27}\,\mathrm{kg}) \times (2.99792 \times 10^{8}\,\mathrm{m/s})^2$$

$$= 1.49242 \times 10^{-10}\,\mathrm{kg}\,(\mathrm{m/s})^2 = 1.49242 \times 10^{-10}\,\mathrm{J}\,(\text{joule}) \times (1\,\mathrm{MeV} / 1.60218 \times 10^{-13}\,\mathrm{J})$$

$$= 931.49\,\mathrm{MeV},$$

so $1\,\mathrm{u}\,c^2 = 931.49\,\mathrm{MeV}$.

Hence, the energy released is 0.0238 × 931 MeV = 22.2 MeV.

Expressed differently: the mass is reduced by 0.3%, corresponding to 0.3% of 90 PJ/kg is 270 TJ/kg.

This is a large amount of energy for a nuclear reaction; the amount is so high because the binding energy per nucleon of the helium-4 nucleus is unusually high, because the He-4 nucleus is "doubly magic". (The He-4 nucleus is unusually stable and tightly bound for the same reason that the helium atom is inert: each pair of protons and neutrons in He-4 occupies a filled 1snuclear orbital in the same way that the pair of electrons in the helium atom occupy a filled 1selectron orbital). Consequently, alpha particles appear frequently on the right hand side of nuclear reactions.

The energy released in a nuclear reaction can appear mainly in one of three ways:

- kinetic energy of the product particles;
- emission of very high energy photons, called gamma rays;
- some energy may remain in the nucleus, as a metastableenergy level.

When the product nucleus is metastable, this is indicated by placing an asterisk ("*") next to its atomic number. This energy is eventually released through nuclear decay.

A small amount of energy may also emerge in the form of X-rays. Generally, the product nucleus has a different atomic number, and thus the configuration of its electron shells is wrong. As the electrons rearrange themselves and drop to lower energy levels, internal transition X-rays (X-rays with precisely defined emission lines) may be emitted.

## Q-value and Energy Balance

In writing down the reaction equation, in a way analogous to a chemical equation, one may in addition give the reaction energy on the right side:

$$\text{Target nucleus} + \text{projectile} \rightarrow \text{Final nucleus} + \text{ejectile} + Q.$$

For the particular case discussed above, the reaction energy has already been calculated as Q = 22.2 MeV. Hence:

$$^{6}_{3}\text{Li} + ^{2}_{1}\text{H} \rightarrow 2^{4}_{2}\text{He} + 22.2\text{MeV}.$$

The reaction energy (the "Q-value") is positive for exothermal reactions and negative for endothermal reactions. On the one hand, it is the difference between the sums of kinetic energies on the final side and on the initial side. But on the other hand, it is also the difference between the nuclear rest masses on the initial side and on the final side (in this way, we have calculated the Q-value above).

## Reaction Rates

If the reaction equation is balanced, that does not mean that the reaction really occurs. The rate at which reactions occur depends on the particle energy, the particle flux and the reaction cross section. An example of a large repository of reaction rates is the REACLIB database, as maintained by the Joint Institute for Nuclear Astrophysics.

## Neutrons vs. Ions

In the initial collision which begins the reaction, the particles must approach closely enough so that the short range strong force can affect them. As most common nuclear particles are positively charged, this means they must overcome considerable electrostatic repulsion before the reaction can begin. Even if the target nucleus is part of a neutral atom, the other particle must penetrate well beyond the electron cloud and closely approach the nucleus, which is positively charged. Thus, such particles must be first accelerated to high energy, for example by:

- particle accelerators;
- nuclear decay (alpha particles are the main type of interest here, since beta and gamma rays are rarely involved in nuclear reactions);
- very high temperatures, on the order of millions of degrees, producing thermonuclear reactions;
- cosmic rays.

Also, since the force of repulsion is proportional to the product of the two charges, reactions between heavy nuclei are rarer, and require higher initiating energy, than those between a heavy and light nucleus; while reactions between two light nuclei are the most common ones.

Neutrons, on the other hand, have no electric charge to cause repulsion, and are able to initiate a nuclear reaction at very low energies. In fact, at extremely low particle energies (corresponding, say, to thermal equilibrium at room temperature), the neutron's de Broglie wavelength is greatly increased, possibly greatly increasing its capture cross section, at energies close to resonances of the nuclei involved. Thus low-energy neutrons *may* be even more reactive than high energy neutrons.

## Notable Types

While the number of possible nuclear reactions is immense, there are several types which are more common, or otherwise notable. Some examples include:

- Fusion reactions — two light nuclei join to form a heavier one, with additional particles (usually protons or neutrons) emitted subsequently.
- Spallation — a nucleus is hit by a particle with sufficient energy and momentum to knock out several small fragments or smash it into many fragments.
- Induced gamma emission belongs to a class in which only photons were involved in creating and destroying states of nuclear excitation.
- Alpha decay — Though driven by the same underlying forces as spontaneous fission, α decay is usually considered to be separate from the latter. The often-quoted idea that "nuclear reactions" are confined to induced processes is incorrect. "Radioactive decays" are a subgroup of "nuclear reactions" that are spontaneous rather than induced. For example, so-called "hot alpha particles" with unusually high energies may actually be produced in induced ternary fission, which is an induced nuclear reaction (contrasting with spontaneous fission). Such alphas occur from spontaneous ternary fission as well.
- Fission reactions — a very heavy nucleus, after absorbing additional light particles (usually neutrons), splits into two or sometimes three pieces. This is an induced nuclear reaction. Spontaneous fission, which occurs without assistance of a neutron, is usually not considered a nuclear reaction. At most, it is not an *induced* nuclear reaction.

## Direct Reactions

An intermediate energy projectile transfers energy or picks up or loses nucleons to the nucleus in a single quick ($10^{-21}$ second) event. Energy and momentum transfer are relatively small. These are particularly useful in experimental nuclear physics, because the reaction mechanisms are often simple enough to calculate with sufficient accuracy to probe the structure of the target nucleus.

## Inelastic Scattering

Only energy and momentum are transferred.

- (p,p') tests differences between nuclear states.

- (α,α') measures nuclear surface shapes and sizes. Since α particles that hit the nucleus react more violently, elastic and shallow inelastic α scattering are sensitive to the shapes and sizes of the targets, like light scattered from a small black object.
- (e,e') is useful for probing the interior structure. Since electrons interact less strongly than do protons and neutrons, they reach to the centers of the targets and their wave functions are less distorted by passing through the nucleus.

## Transfer Reactions

Usually at moderately low energy, one or more nucleons are transferred between the projectile and target. These are useful in studying outer shell structure of nuclei.

- (α,n) and (α,p) reactions. Some of the earliest nuclear reactions studied involved an alpha particle produced by alpha decay, knocking a nucleon from a target nucleus.
- (d,n) and (d,p) reactions. A deuteronbeam impinges on a target; the target nuclei absorb either the neutron or proton from the deuteron. The deuteron is so loosely bound that this is almost the same as proton or neutron capture. A compound nucleus may be formed, leading to additional neutrons being emitted more slowly. (d,n) reactions are used to generate energetic neutrons.
- The strangeness exchange reaction (K, π) has been used to study hypernuclei.
- The reaction $^{14}N(\alpha,p)^{17}O$ performed by Rutherford in 1917 (reported 1919), is generally regarded as the first nuclear transmutation experiment.

## Reactions with Neutrons

| | → T | → $^{7}Li$ | → $^{14}C$ | | |
|---|---|---|---|---|---|
| (n,α) | $^{6}Li + n \rightarrow T + \alpha$ | $^{10}B + n \rightarrow {}^{7}Li + \alpha$ | $^{17}O + n \rightarrow {}^{14}C + \alpha$ | $^{21}Ne + n \rightarrow {}^{18}O + \alpha$ | $^{37}Ar + n \rightarrow {}^{34}S + \alpha$ |
| (n,p) | $^{3}He + n \rightarrow T + p$ | $^{7}Be + n \rightarrow {}^{7}Li + p$ | $^{14}N + n \rightarrow {}^{14}C + p$ | $^{22}Na + n \rightarrow {}^{22}Ne + p$ | |

Reactions with neutrons are important in nuclear reactors and nuclear weapons. While the best-known neutron reactions are neutron scattering, neutron capture, and nuclear fission, for some light nuclei (especially odd-odd nuclei) the most probable reaction with a thermal neutron is a transfer reaction:

Some reactions are only possible with fast neutrons:

- (n,2n) reactions produce small amounts of protactinium-231 and uranium-232 in the thorium cycle which is otherwise relatively free of highly radioactive actinide products.
- $^{9}Be + n \rightarrow 2\alpha + 2n$ can contribute some additional neutrons in the berylliumneutron reflector of a nuclear weapon.
- $^{7}Li + n \rightarrow T + \alpha + n$ unexpectedly contributed additional yield in Castle Bravo, Castle Romeo, and Castle Yankee, the three highest-yield nuclear tests conducted by the U.S.

### Compound Nuclear Reactions

Either a low-energy projectile is absorbed or a higher energy particle transfers energy to the nucleus, leaving it with too much energy to be fully bound together. On a time scale of about $10^{-19}$ seconds, particles, usually neutrons, are "boiled" off. That is, it remains together until enough energy happens to be concentrated in one neutron to escape the mutual attraction. Charged particles rarely boil off because of the coulomb barrier. The excited quasi-bound nucleus is called a compound nucleus.

- Low energy (e, e' xn), (γ, xn) (the xn indicating one or more neutrons), where the gamma or virtual gamma energy is near the giant dipole resonance. These increase the need for radiation shielding around electron accelerators.

## Q Value (Nuclear Science)

In nuclear physics and chemistry, the *Q* value for a reaction is the amount of energy released by that reaction. The value relates to the enthalpy of a chemical reaction or the energy of radioactive decay products. It can be determined from the masses of reactants and products. Q values affect reaction rates. In general, the larger the positive Q value for the reaction, the faster the reaction proceeds, and the more likely the reaction is to "favor" the products.

### Definition

Considering the energy conservation of the simple reaction, enables the general definition of *Q* based on mass-energy equivalence, where *K* is kinetic energy and *m* is mass:

$$Q = K_{\mathrm{f}} - K_{\mathrm{i}} = (m_{\mathrm{i}} - m_{\mathrm{f}})c^2$$

A reaction with a positive *Q* value is exothermic, i.e. has a net release of energy, since the kinetic energy of the final state is greater than the kinetic energy of the initial state. A reaction with a negative *Q* value is endothermic, i.e. requires a net energy input, since the kinetic energy of the final state is less than the kinetic energy of the initial state.

### Applications

Chemical *Q* values are measurement in calorimetry. Exothermic chemical reactions tend to be more spontaneous and can emit light or heat, resulting in runaway feedback(i.e. explosions).

*Q* values are also featured in particle physics. For example, Sargent's rule states that weak reaction rates are proportional to $Q^5$. The *Q* value is the kinetic energy released in the decay at rest. For neutron decay, some mass disappears as neutrons convert to a proton, electron and antineutrino:

$$Q = (m_{\mathrm{n}} - m_{\mathrm{p}} - m_{\bar{\mathrm{v}}} - m_{\mathrm{e}})c^2 = K_{\mathrm{p}} + K_{\mathrm{e}} + \mathrm{K}_{\bar{\mathrm{v}}} = 0.782 \text{ MeV}$$

where $m_n$ is the mass of the neutron, $m_p$ is the mass of the proton, $m_v$ is the mass of the electron antineutrino and $m_e$ is the mass of the electron; and the $K$ are the corresponding kinetic energies. The neutron has no initial kinetic energy since it is at rest. In beta decay, a typical $Q$ is around 1 MeV.

The decay energy is divided among the products in a continuous distribution for more than 2 products. Measuring this spectrum allows one to find the mass of a product. Experiments are studying emission spectrums to search for neutrinoless decay and neutrino mass; this is the principle of the upcoming KATRIN experiment.

## Energetic of Nuclear Reaction - Q Value

The first and foremost point of nuclear reaction is to understand whether a nuclear reaction yields energy or just has a reverse effect. Therefore, it is important to calculate energy gain or loss in a nuclear reaction in terms of its Q-value. Let us first calculate it from the following consideration:

When a target nucleus is bombarded with an energetic projectile a different nucleus forms along with another particle. In this process either energy gets released or absorbed depending upon how masses of the reactants change, and is known as nuclear reaction. Nuclear physicists understood this process on the basis of formation of a 'compound nucleus' when a projectile interacts with a target nucleus. The compound nucleus can decay in different channels forming various product nuclei and other particles and does not remember the way it was formed. The application of nuclear reaction is immense. The two most important nuclear reactions, i.e. fission and fusion are the most efficient ways of electrical power production and hence are part and parcel of human civilization.

Let us first discuss what the Q-value of a nuclear reaction is and how to calculate it.

In general, any nuclear reaction can be represented as

$$a + A = B + b$$

where,

A = target nucleus with mass MA

B = Product nucleus of mass MB

a = incident particle or projectile of mass ma

b = product particle of mass mb

Let us take $k_A, k_B, k_a, k_b$ be the kinetic energies of A, B, a and b respectively.

From conservation of energy principle, we can write

$$(\text{K.E.} + \text{Rest mass energy})\text{L.H.S.} = (\text{K.E.} + \text{Rest mass energy})\text{R.H.S}$$

$$\text{or, } M_A c^2 + \left(k_a + m_a c^2\right) = \left(k_B + M_B c^2\right) + \left(k_b + m_b c^2\right)$$

Considering target nucleus is at rest i.e., $k_A = 0$.

$$\text{or, } k_B + k_b - k_a = (M_A + m_a - M_B - m_a)c^2$$

where, c is the velocity of light in vacuum.

The difference in K.E. of reaction products and projectile is known as the energy balance or Q-value of the reaction.

$$\therefore Q = k_B + k_b - k_a = (M_A + m_a - M_B - m_a)c^2$$

Depending on whether Q-value is positive or negative, the nuclear reactions can be classified into two categories; namely, exoergic and endoergic. In exoergic reaction, (MA +ma) > (MB +mb), i.e. (kB +kb) >ka and hence Q > 0. When the input mass is greater than the output mass, some mass is lost in the form of energy at the expense of lost mass. On the other hand, in endoergic reaction Q < 0, i.e. the output mass being larger than the input mass, some mass has been created at the expense or 'loss' of the output kinetic energy.

Let us see the following example to understand how to calculate Q-value of a typical nuclear reaction.

Example m3.1: The masses of the different nuclei taking part in ${}_3^1Li(p,n)$ reaction in amu are as follows: $M({}_3^1Li) = 7.01822, M({}_1^1H) = 1.00814, M({}_0^1n) = 1.00898$ and mass of the product nucleus = 7.01915. Calculate the Q-value of this reaction in MeV. Is it exoergic or endoergeic?

Solution: Masses of the reactants = 7.01822 + 1.00814 = 8.02636 amu Masses of the products = 7.01915 + 1.0898 = 8.02813 amu Q-value of the reaction = 8.02636-8.02813 = -0.00117 amu = -0.00117 X 931 MeV = -1.089 MeV.

The negative sign of the Q-value indicates that the reaction is endoergic.

Now we will see the physical quantities those remain conserved during a nuclear reaction. You know that when two billiard balls hit each other, their motion after collision is strictly decided by the conservation of energy and momentum. Similarly, in all physical processes some quantities must be conserved.

In nuclear reaction the following quantities are conserved:

(a) Conservation of energy (including mass-energy)

(b) Conservation of linear momentum

(c) Conservation of angular momentum (spin and orbital)

(d) Conservation of electric charge

(e) Conservation of statistics and

(f) Conservation of parity (except in weak interaction).

Let me just explain the conservation of last two quantities, i.e. statistics and parity. It is known that Fermi-Dirac (F-D) statistics is applied to a particle of half-integral spin (e.g. a nucleon) or a group of nucleons of odd number. When the number of nucleons is even, Bose-Einstein (B-E) statistics is applied, which is otherwise applicable to particles of integral spin. Let us denote F-D statistics by -1 and B-E statistics by $+1^1$. The simplest example of conservation of statistics can be shown in electron-positron annihilation

$$e^- + e^+ \rightarrow y + y$$

The statistics in the left hand side of the above reaction is (-1) X (-1) = +1. Annihilation process gives rise two gamma photons, which are bosons and hence the statistics in the right hand side is (+1) X (+1) = +1. Parity is a purely mathematical concept, which describes the kind of spatial symmetry of physical phenomena. If the sign of the wave function ψ may or may not change in sign with reflection symmetry (i.e. $\vec{r} \rightarrow -\vec{r}$). First case it is known as even parity and the second case the odd parity. Therefore,

$$\Psi\left(\vec{r}\right) = +\Psi\left(-\vec{r}\right) \qquad \text{Even parity}$$

$$\Psi\left(\vec{r}\right) = -\Psi\left(-\vec{r}\right) \qquad \text{Odd parity}$$

The product of the parities of individual particle gives the parity of a system of particles. All the elementary particle and nuclear energy state have either even or odd parity. Apart from nuclear reactions involving weak interaction, parity is conserved in all other nuclear reactions.

However, in weak interaction a quantity combining charge and parity remains We know that nucleons being spin ½ particles, satisfy Fermi-Dirac statistics.

Let us now address the question:

How nuclear reaction takes place? -Compound nucleus formation

For relatively lower projectile energies, reactions may not involve a direct mechanism, but may proceed via an unstable intermediate state known as compound nucleus state. In this case, the incident particle loses so much energy that it cannot escape the struck nucleus. The compound nucleus thus formed has its mass number equal to the mass numbers of the target nucleus and that of the projectile. Compound nucleus is in highly excited state and therefore breaks into a product nucleus and other particles. Classically, compound nucleus can be understood by this qualitative manner. According to liquid drop model, nucleus can be considered to be a conglomeration of particles held together by attractive forces acting between them. An energetic nucleon approaching this nucleus enters into the target nucleus under an attractive potential and probably collides with one of the nucleons, and the energy is shared between both of them. Each nucleon again has the probability of colliding with other nucleons and thereby sharing the energy and ultimately shared between many nucleons. In this situation, it is called a compound nucleus. A typical example of nuclear reaction through compound nuclear formation is

$$Stage\ 1(formation) \qquad {}_1^1H + {}_{13}^{27}Al \rightarrow {}_{14}^{28}S_i^*$$

$$Stage \quad (decay) \qquad {}^{28}_{14}S^* \rightarrow {}^{24}_{12}Mg + {}^{4}_{2}He$$

One important characteristics of compound nucleus formation theory is that, compound nucleus may be formed in a number of different ways; however, the properties of compound nucleus are independent of the mode of formation.

One important concept in the compound nucleus picture of nuclear reaction is the 'level width'. The compound nucleus, on an average, remains in a given excited state for a certain time before decaying either through particle or γ-ray emission. The reciprocal of this mean life time, τ, is called the decay constant which represents the probability per unit time of the emission of a particle or γ-ray. The level width, Γ is defined as

$$\Gamma = \frac{\hbar}{T} = \hbar\lambda$$

where, λ is the decay constant. From energy-time uncertainty principle,

$$\Delta E \Delta t \approx \hbar$$

considering Δt to be the time uncertainty corresponding to the energy uncertainty ΔE, to be the mean life time of an excited state, the level width Γ is then identified with ΔE. In nuclear reaction, different decay modes posses different level widths. The total width may be written as the sum of the partial widths

$$\Gamma = \Gamma_a + \Gamma_\alpha + \Gamma_n + \Gamma_y$$

where $\Gamma_d, \Gamma_\alpha, \Gamma_n, \Gamma_y$ etc. are the partial level widths for the emission of deuteron, α-particles, neutrons, γ-rays etc. The concept of level width is useful in that it can be obtained from the measurement of resonances. By knowing the total level width, the mean life time can be calculated from the relation

$$\tau(\sec) = \frac{\lambda(joule - \sec)}{\Gamma(joule)} = \frac{1.106\times10^{-34}}{\Gamma(eV)\times1.6\times10^{-19}} = \frac{6.9\times10^{-16}}{\Gamma(eV)}$$

Following example illustrates mean life of a compound nucleus.

Example m3.2: The reaction shows among others a resonance for all excitation energy of the compound nucleus of 13.23 MeV. The width of this level as found experimentally is 130 keV. Calculate the mean life of the nucleus for this excitation.1

Solution: The reaction is

$$B_5^{10} + H_{e2}^{4} \rightarrow \left(N_7^{14}\right)^* \rightarrow C_6^{13} + H_1^1$$

The mean lifetime is given by

$$\tau = \frac{6.9\times10^{-16}}{13\times10^4} \approx 5\times10^{-21}\sec$$

## Reaction Cross Section and Examples of Various Nuclear Reactions

As nuclear reaction is a statistical phenomenon, it is required to define some physical quantity to determine the probability of a nuclear reaction. The quantity which gives the idea of the probability of any physical process (e.g. nuclear reaction) to occur is known as cross-section. Nuclear reaction cross section can be defined in the following manner:

σ = Number of given types of events per unit time per nucleus/number of projectile particles per unit area, unit time.

Considering two broad physical process i.e. scattering and absorption, the total cross section σtot is written as:

$$\sigma_{tot} = \sigma_{sc} + \sigma_a$$

whereσsc = scattering cross-section

andσa = absorption cross-section

The unit of cross-section is 'barn' having the dimension of area $\left(1\ barn\ =\ 10^{-24} cm^2 = 10^{-28} m^2\right)$.

Now we will see examples of various nuclear reactions:

A nuclear reaction is symbolized by a parenthesis containing the projectile and product particle symbolically. At the beginning of the parenthesis, the symbol of target nucleus, and after the parenthesis, the symbol of the product nucleus is written. To represent a particular reaction, say, a deuteron irradiating to produce and a proton, the symbol is . Depending on the projectile and the product particle, a large number of nuclear reactions are possible.

### (d, α) Reactions

These reactions yield α particles, while projectile is deuteron. For example,

$$^{6}_{3}Li + ^{2}_{1}H \rightarrow \left\lfloor ^{8}_{4}Be \right\rfloor \rightarrow ^{4}_{2}He + ^{4}_{2}He$$

$$^{27}_{13}Al + ^{2}_{1}H \rightarrow \left\lfloor ^{29}_{14}Si \right\rfloor \rightarrow ^{25}_{12}Mg + ^{4}_{2}He$$

The Q values of these reactions are positive and the reactions are exoergic.

### (d, p) Reactions

These reactions yield protons. For example,

$$^{6}_{3}Li + ^{2}_{1}H \rightarrow \left\lfloor ^{8}_{4}Be \right\rfloor \rightarrow ^{7}_{4}Li + ^{1}_{1}H$$

$$^{27}_{13}Na + ^{2}_{1}H \rightarrow \left\lfloor ^{25}_{12}Mg \right\rfloor \rightarrow ^{24}_{11}Na + ^{1}_{1}H \cdot$$

The Q-values are usually positive and the reactions are exoergic.

## (d,n) Reactions

These reactions yield neutrons. For example,

$$ {}_{4}^{9}Be + {}_{1}^{2}H \rightarrow \left\lfloor {}_{5}^{11}B \right\rfloor \rightarrow {}_{5}^{10}B + {}_{0}^{1}n \cdot $$

$$ {}_{47}^{107}Ag + {}_{1}^{2}H \rightarrow \left\lfloor {}_{48}^{109}Cd \right\rfloor \rightarrow {}_{48}^{107}Cd + {}_{0}^{1}n + {}_{0}^{1}n $$

The reaction (4.21) has the importance, as it serves as sources of neutrons.

## (α, p) Reactions

These reactions yield protons while projectile is α particle. For example

$$ {}_{7}^{14}N + {}_{2}^{4}He \rightarrow \left\lfloor {}_{9}^{18}F \right\rfloor \rightarrow {}_{8}^{17}O + {}_{1}^{1}H $$

$$ {}_{11}^{23}Na + {}_{2}^{4}He \rightarrow \left\lfloor {}_{13}^{27}Al \right\rfloor \rightarrow {}_{12}^{26}Mg + {}_{1}^{1}H \cdot $$

## (α, n) Reactions

These reactions yield neutrons and also utilized as convenient source for neutrons. For example,

$$ {}_{4}^{9}Be + {}_{2}^{4}He \rightarrow \left\lfloor {}_{6}^{13}C \right\rfloor \rightarrow {}_{12}^{26}C + {}_{0}^{1}n $$

$$ {}_{3}^{7}Li + {}_{2}^{4}He \rightarrow \left\lfloor {}_{5}^{11}B \right\rfloor \rightarrow {}_{5}^{10}B + {}_{0}^{1}n $$

For reaction (4.25), the α-particles is supplied by radium and its products and beryllium provide an inexpensive source for neutrons.

## (α, γ) Reactions

If a light element such as Li is bombarded with α particles of variable energy, it is observed that at certain values of energy, the compound nucleus de-excites by γ-emission. The process is called radiative capture of α-particles. For example,

$$ {}_{3}^{7}Li + {}_{2}^{4}He \rightarrow \left\lfloor {}_{5}^{11}B \right\rfloor \rightarrow {}_{5}^{11}B + y $$

## (p, α) Reactions

These reactions yield α-particles with protons as projectile. For example,

$$ {}_{3}^{7}Li + {}_{1}^{1}H \rightarrow \left\lfloor {}_{4}^{8}Be \right\rfloor \rightarrow {}_{2}^{4}He + {}_{2}^{4}He $$

which is an exoergic reaction. Another example of this reaction is

$$^{10}_{5}B + ^{1}_{1}H \rightarrow \left[^{11}_{6}C\right] \rightarrow ^{7}_{4}Be + ^{4}_{2}He$$

γ - Induced Transmutationss

Target nucleus bombarded with γ-radiation usually results in a compound nucleus in an excited state. The excess excitation energy is released by the nuclei through neutron emission of different particles. Some typical examples are

$$^{2}_{1}H + \gamma \rightarrow \left[^{2}_{1}H^{*}\right] \rightarrow ^{1}_{1}H + ^{1}_{0}n\,(\gamma, n)\ reaction$$

$$^{12}_{6}C + \gamma \rightarrow \left[^{12}_{6}C^{*}\right] \rightarrow ^{8}_{4}Be + ^{4}_{2}He\,(\gamma, \alpha)\ reaction$$

$$^{10}_{5}B + \gamma \rightarrow \left[^{10}_{5}B^{*}\right] \rightarrow ^{8}_{4}Be + ^{2}_{1}H\,(\gamma, \mathrm{d})\ reaction$$

## Neutron Induced Transmutations

Depending on the energy, neutrons are classified as fast and slow (discussed in chapter 2). Slow neutrons with thermal equilibrium with the medium through which they pass are termed as thermal neutrons. When neutrons strike a target, they may either be scattered or captured by the target nuclei to form compound nuclei, which eventually emit γ-photons, α-particles, deuterons etc. Some typical examples of neutron induced transmutations are:

$$^{6}_{3}Li + ^{1}_{0}n \rightarrow \left[^{7}_{3}Li\right] \rightarrow ^{3}_{1}H + ^{4}_{2}He\,(\gamma, \alpha)\ reaction$$

This reaction is useful to produce tritium, which is useful in nuclear fusion.

$$^{14}_{7}N + ^{1}_{0}n \rightarrow \left[^{15}_{7}N\right] \rightarrow ^{14}_{6}C + ^{1}_{1}H \text{, followed by}$$

$$^{14}_{7}C \rightarrow ^{14}_{7}N + e^{-} + \bar{\nu}_{e}$$

which is an example of (n,p) reaction. These reactions finally produce the same target nucleus. Almost all elements exhibit radiative capture of low energy neutrons. This capture raises the target nucleus to an excited isomeric state. This excitation energy is subsequently released in the form of γ-photons and the product nucleus becomes an istope of the original nucleus. For example,

$$^{27}_{13}Al + ^{1}_{0}n \rightarrow \left[^{28}_{13}Al\right]^{*} \rightarrow ^{28}_{13}Al + \gamma\,(n, \gamma)\,reaction.$$

Now we will see two most important nuclear reactions as far as energy production is concerned known as nuclear fission and fusion reaction.

## Nuclear Fission Reaction

In 1939, Otto Hahn and Strassmann discovered a very important nuclear reaction where  was bombarded with slow neutrons forming a compound nucleus, which eventually got fragmented into two nuclei and further neutrons were produced. This is known as fission reaction and is represented by:

$$ {}^{235}_{92}U + {}^{1}_{0}n \rightarrow \left\lfloor {}^{236}_{92}U \right\rfloor \rightarrow X + Y + neutrons $$

where ${}^{236}_{92}U$ is a highly unstable isotope and X ,Y are the fission fragments. There could be various combinations of the fission fragments and that a number of neutrons are also given off. Typical fission reactions are

$$ {}^{1}_{0}n + {}^{235}_{92}U \rightarrow \left\lfloor {}^{236}_{92}U \right\rfloor \rightarrow {}^{144}_{56}Ba + \frac{89}{36}Kr + 3{}^{1}_{0}n + Q $$

$$ {}^{1}_{0}n + {}^{235}_{92}U \rightarrow \left\lfloor {}^{236}_{92}U \right\rfloor \rightarrow {}^{144}_{54}Xe + \frac{94}{38}Sr + 2{}^{1}_{0}n + Q ------- \quad m3.34 $$

where, Q is the energy released in the reaction. The detail of energy release in nuclear fission reaction and its importance in electrical power generation.

## Nuclear Fusion Reaction

Another important nuclear reaction is the nuclear fusion reaction. In this reaction, light nuclei combine or fuse together to produce a relatively heavy nucleus, there would be a greater binding energy and consequent decrease in nuclear mass resulting in a positive Q-value of the reaction. Example of a typical fusion reaction is

$$ {}^{1}_{1}H + {}^{1}_{1}H \rightarrow {}^{2}_{1}H + {}^{0}_{1}e + \upsilon + Q $$

$$ {}^{2}_{1}H + {}^{2}_{1}H \rightarrow {}^{3}_{2}He + {}^{1}_{0}n + Q $$

The importance of the fusion reaction in the stellar energy production was first suggested by Hans Bethe in 1938. In the next chapter, we will discuss the energy release of this equation and its importance in future power generation process.

# Nuclear Reactor

Core of CROCUS, a small nuclear reactor used for research at the EPFL in Switzerland

A nuclear reactor, formerly known as an atomic pile, is a device used to initiate and control a sustained nuclear chain reaction. Nuclear reactors are used at nuclear power plants for electricity generation and in propulsion of ships. Heat from nuclear fission is passed to a working fluid (water or gas), which runs through steam turbines. These either drive a ship's propellers or turn electrical generators. Nuclear generated steam in principle can be used for industrial process heat or for district heating. Some reactors are used to produce isotopes for medical and industrial use, or for production of weapons-gradeplutonium. Some are run only for research. As of April 2014, the IAEA reports there are 435 nuclear power reactors in operation, in 31 countries around the world.

## Mechanism

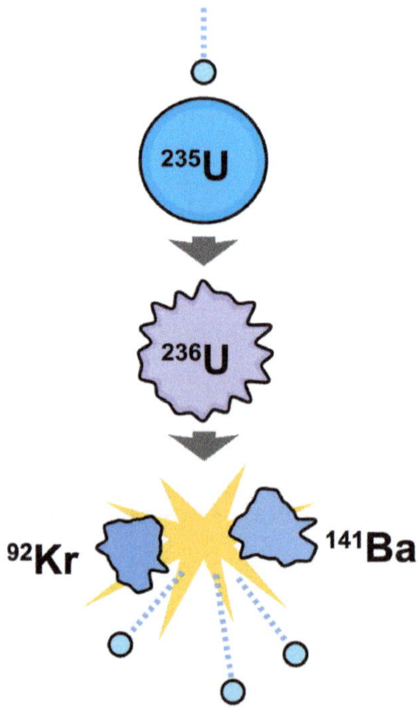


An induced nuclear fission event. A neutron is absorbed by the nucleus of a uranium-235 atom, which in turn splits into fast-moving lighter elements (fission products) and free neutrons. Though both reactors and nuclear weapons rely on nuclear chain-reactions, the rate of reactions in a reactor occurs much more slowly than in a bomb.

Just as conventional power-stations generate electricity by harnessing the thermal energy released from burning fossil fuels, nuclear reactors convert the energy released by controlled nuclear fission into thermal energy for further conversion to mechanical or electrical forms.

## Fission

When a large fissileatomic nucleus such as uranium-235 or plutonium-239 absorbs a neutron, it may undergo nuclear fission. The heavy nucleus splits into two or more lighter nuclei, (the fission products), releasing kinetic energy, gamma radiation, and free neutrons. A portion of these neutrons may later be absorbed by other fissile atoms and trigger further fission events, which release more neutrons, and so on. This is known as a nuclear chain reaction.

To control such a nuclear chain reaction, neutron poisons and neutron moderators can change the portion of neutrons that will go on to cause more fission. Nuclear reactors generally have automatic and manual systems to shut the fission reaction down if monitoring detects unsafe conditions.

Commonly-used moderators include regular (light) water (in 74.8% of the world's reactors), solid graphite (20% of reactors) and heavy water (5% of reactors). Some experimental types of reactor have used beryllium, and hydrocarbons have been suggested as another possibility.

## Heat Generation

The reactor core generates heat in a number of ways:

- The kinetic energy of fission products is converted to thermal energy when these nuclei collide with nearby atoms.
- The reactor absorbs some of the gamma rays produced during fission and converts their energy into heat.
- Heat is produced by the radioactive decay of fission products and materials that have been activated by neutron absorption. This decay heat-source will remain for some time even after the reactor is shut down.

A kilogram of uranium-235 (U-235) converted via nuclear processes releases approximately three million times more energy than a kilogram of coal burned conventionally ($7.2 \times 10^{13}$ joules per kilogram of uranium-235 versus $2.4 \times 10^{7}$ joules per kilogram of coal).

## Cooling

A nuclear reactor coolant — usually water but sometimes a gas or a liquid metal (like liquid sodium) or molten salt — is circulated past the reactor core to absorb the heat that it generates. The heat is carried away from the reactor and is then used to generate steam. Most reactor systems employ a cooling system that is physically separated from the water that will be boiled to produce pressurized steam for the turbines, like the pressurized water reactor. However, in some reactors the water for the steam turbines is boiled directly by the reactor core; for example the boiling water reactor.

## Reactivity Control

The power output of the reactor is adjusted by controlling how many neutrons are able to create more fissions.

Control rods that are made of a neutron poison are used to absorb neutrons. Absorbing more neutrons in a control rod means that there are fewer neutrons available to cause fission, so pushing the control rod deeper into the reactor will reduce its power output, and extracting the control rod will increase it.

At the first level of control in all nuclear reactors, a process of delayed neutron emission by a number of neutron-rich fission isotopes is an important physical process. These delayed neutrons account for about 0.65% of the total neutrons produced in fission, with the remainder (termed "prompt neutrons") released immediately upon fission. The fission products which produce delayed neutrons have half lives for their decay by neutron emission that range from milliseconds to as long as several minutes, and so considerable time is required to determine exactly when a reactor reaches the critical point. Keeping the reactor in the zone of chain-reactivity where delayed neutrons are *necessary* to achieve a critical mass state allows mechanical devices or human operators to control a chain reaction in "real time"; otherwise the time between achievement of criticality and nuclear meltdown as a result of an exponential power surge from the normal nuclear chain reaction, would be too short to allow for intervention. This last stage, where delayed neutrons are no longer required to maintain criticality, is known as the prompt critical point. There is a scale for describing criticality in numerical form, in which bare criticality is known as *zero dollars* and the prompt critical point is *one dollar*, and other points in the process interpolated in cents.

In some reactors, the coolant also acts as a neutron moderator. A moderator increases the power of the reactor by causing the fast neutrons that are released from fission to lose energy and become thermal neutrons. Thermal neutrons are more likely than fast neutrons to cause fission. If the coolant is a moderator, then temperature changes can affect the density of the coolant/moderator and therefore change power output. A higher temperature coolant would be less dense, and therefore a less effective moderator.

In other reactors the coolant acts as a poison by absorbing neutrons in the same way that the control rods do. In these reactors power output can be increased by heating the coolant, which makes it a less dense poison. Nuclear reactors generally have automatic and manual systems to scram the reactor in an emergency shut down. These systems insert large amounts of poison (often boron in the form of boric acid) into the reactor to shut the fission reaction down if unsafe conditions are detected or anticipated.

Most types of reactors are sensitive to a process variously known as xenon poisoning, or the iodine pit. The common fission productXenon-135 produced in the fission process acts as a neutron poison that absorbs neutrons and therefore tends to shut the reactor down. Xenon-135 accumulation can be controlled by keeping power levels high enough to destroy it by neutron absorption as fast as it is produced. Fission also produces iodine-135, which in turn decays (with a half-life of 6.57 hours) to new xenon-135. When the reactor is shut down, iodine-135 continues to decay to xenon-135, making restarting the reactor more difficult for a day or two, as the xenon-135 decays into cesium-135, which is not nearly as poisonous as xenon-135, with a half-life of 9.2 hours. This temporary state is the "iodine pit." If the reactor has sufficient extra reactivity capacity, it can be restarted. As the extra xenon-135 is transmuted to xenon-136, which is much less a neutron poison, within a few hours the reactor experiences a "xenon burnoff (power) transient". Control rods must be further inserted to replace the neutron absorption of the lost xenon-135. Failure to properly follow such a procedure was a key step in the Chernobyl disaster.

Reactors used in nuclear marine propulsion (especially nuclear submarines) often cannot be run at continuous power around the clock in the same way that land-based power reactors are normally run, and in addition often need to have a very long core life without refueling. For this reason many designs use highly enriched uranium but incorporate burnable neutron poison in the fuel rods. This allows the reactor to be constructed with an excess of fissionable material, which is nevertheless made relatively safe early in the reactor's fuel burn-cycle by the presence of the neutron-absorbing material which is later replaced by normally produced long-lived neutron poisons (far longer-lived than xenon-135) which gradually accumulate over the fuel load's operating life.

## Electrical Power Generation

The energy released in the fission process generates heat, some of which can be converted into usable energy. A common method of harnessing this thermal energy is to use it to boil water to produce pressurized steam which will then drive a steam turbine that turns an alternator and generates electricity.

## Early Reactors

The Chicago Pile, the first nuclear reactor, built in secrecy at the University of Chicago in 1942 during World War II as part of the US's Manhattan project.

The neutron was discovered in 1932. The concept of a nuclear chain reaction brought about by nuclear reactions mediated by neutrons was first realized shortly thereafter, by Hungarian scientist LeóSzilárd, in 1933. He filed a patent for his idea of a simple reactor the following year while working at the Admiralty in London. However, Szilárd's idea did not incorporate the idea of nuclear fission as a neutron source, since that process was not yet discovered. Szilárd's ideas for nuclear reactors using neutron-mediated nuclear chain reactions in light elements proved unworkable.

Lise Meitner and Otto Hahn in their laboratory.

Inspiration for a new type of reactor using uranium came from the discovery by Lise Meitner, Fritz Strassmann and Otto Hahn in 1938 that bombardment of uranium with neutrons (provided by an alpha-on-beryllium fusion reaction, a "neutron howitzer") produced a barium residue, which they reasoned was created by the fissioning of the uranium nuclei. Subsequent studies in early 1939 (one of them by Szilárd and Fermi) revealed that several neutrons were also released during the fissioning, making available the opportunity for the nuclear chain reaction that Szilárd had envisioned six years previously.

On 2 August 1939 Albert Einstein signed a letter to President Franklin D. Roosevelt (written by Szilárd) suggesting that the discovery of uranium's fission could lead to the development of "extremely powerful bombs of a new type", giving impetus to the study of reactors and fission. Szilárd and Einstein knew each other well and had worked together years previously, but Einstein had never thought about this possibility for nuclear energy until Szilard reported it to him, at the beginning of his quest to produce the Einstein-Szilárd letter to alert the U.S. government.

Shortly after, Hitler's Germany invaded Poland in 1939, starting World War II in Europe. The U.S. was not yet officially at war, but in October, when the Einstein-Szilárd letter was delivered to him, Roosevelt commented that the purpose of doing the research was to make sure "the Nazis don't blow us up." The U.S. nuclear project followed, although with some delay as there remained skepticism (some of it from Fermi) and also little action from the small number of officials in the government who were initially charged with moving the project forward.

The following year the U.S. Government received the Frisch–Peierls memorandum from the UK, which stated that the amount of uranium needed for a chain reaction was far lower than had previously been thought. The memorandum was a product of the MAUD Committee, which was working on the UK atomic bomb project, known as Tube Alloys, later to be subsumed within the Manhattan Project.

The Chicago Pile Team, including Enrico Fermi and LeóSzilárd.

Eventually, the first artificial nuclear reactor, Chicago Pile-1, was constructed at the University of Chicago, by a team led by Enrico Fermi, in late 1942. By this time, the program had been pressured for a year by U.S. entry into the war. The Chicago Pile achieved criticality on 2 December 1942 at 3:25 PM. The reactor support structure was made of wood, which supported a pile (hence the name) of graphite blocks, embedded in which was natural uranium-oxide 'pseudospheres' or 'briquettes'.

Soon after the Chicago Pile, the U.S. military developed a number of nuclear reactors for the Manhattan Project starting in 1943. The primary purpose for the largest reactors (located at the Hanford Site in Washington state), was the mass production of plutonium for nuclear weapons. Fermi and Szilard applied for a patent on reactors on 19 December 1944. Its issuance was delayed for 10 years because of wartime secrecy.

"World's first nuclear power plant" is the claim made by signs at the site of the EBR-I, which is now a museum near Arco, Idaho. Originally called "Chicago Pile-4", it was carried out under the direction of Walter Zinn for Argonne National Laboratory. This experimental LMFBR operated by the U.S. Atomic Energy Commission produced 0.8 kW in a test on 20 December 1951 and 100 kW (electrical) the following day, having a design output of 200 kW (electrical).

Besides the military uses of nuclear reactors, there were political reasons to pursue civilian use of atomic energy. U.S. President Dwight Eisenhower made his famous Atoms for Peace speech to the UN General Assembly on 8 December 1953. This diplomacy led to the dissemination of reactor technology to U.S. institutions and worldwide.

The first nuclear power plant built for civil purposes was the AM-1 Obninsk Nuclear Power Plant, launched on 27 June 1954 in the Soviet Union. It produced around 5 MW (electrical).

After World War II, the U.S. military sought other uses for nuclear reactor technology. Research by the Army and the Air Force never came to fruition; however, the U.S. Navy succeeded when they steamed the USS *Nautilus* (SSN-571) on nuclear power 17 January 1955.

The first commercial nuclear power station, Calder Hall in Sellafield, England was opened in 1956 with an initial capacity of 50 MW (later 200 MW).

The first portable nuclear reactor "Alco PM-2A" used to generate electrical power (2 MW) for Camp Century from 1960.

## Components

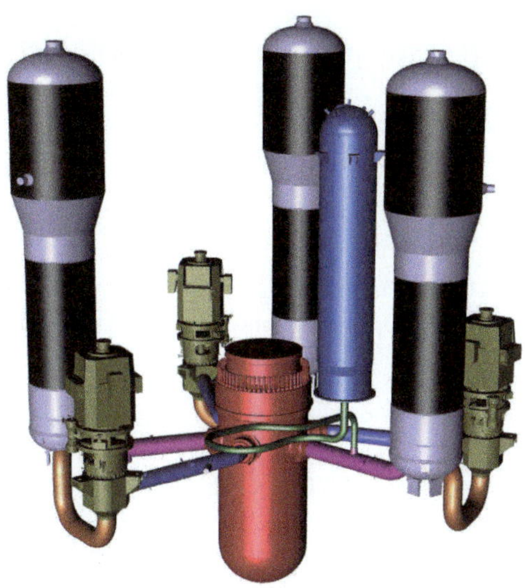

Primary coolant system showing reactor pressure vessel (red), steam generators (purple), pressurizer (blue), and pumps (green) in the three coolant loop Hualong Onepressurized water reactor design

The key components common to most types of nuclear power plants are:

- Nuclear fuel
- Nuclear reactor core
- Neutron moderator
- Startup neutron source
- Neutron poison
- Neutron howitzer (provides steady source of neutrons to re-initiate reaction following shutdown)
- Coolant (often the Neutron Moderator and the Coolant are the same, usually both purified water)
- Control rods
- Reactor pressure vessel (RPV)
- Boiler feedwater pump
- Steam generators (not in BWRs)
- Steam turbine
- Electrical generator
- Condenser
- Cooling tower (not always required)
- Radwaste System (a section of the plant handling radioactive waste)
- Refueling Floor
- Spent fuel pool
- Nuclear safety systems
    - Reactor Protective System (RPS)
    - Emergency Diesel Generators
    - Emergency Core Cooling Systems (ECCS)
    - Standby Liquid Control System (emergency boron injection, in BWRs only)
- Essential service water system (ESWS)

- Containment building
- Control room
- Emergency Operations Facility
- Nuclear training facility (usually contains a Control Room simulator).

## Reactor Types

NC State's PULSTAR Reactor is a 1 MW pool-type research reactor with 4% enriched, pin-type fuel consisting of $UO_2$ pellets in zircaloy cladding.

## Classifications

Nuclear Reactors are classified by several methods; a brief outline of these classification methods is provided.

## Classification by Type of Nuclear Reaction

### Nuclear Fission

All commercial power reactors are based on nuclear fission. They generally use uranium and its product plutonium as nuclear fuel, though a thorium fuel cycle is also possible. Fission reactors can be divided roughly into two classes, depending on the energy of the neutrons that sustain the fission chain reaction:

- Thermal reactors (the most common type of nuclear reactor) use slowed or thermal neutrons to keep up the fission of their fuel. Almost all current reactors are of this type. These contain neutron moderator materials that slow neutrons until their neutron temperature is *thermalized*, that is, until their kinetic energy approaches the average kinetic energy of the surrounding particles. Thermal neutrons have a far higher cross-section (probability) of fissioning the fissile nuclei uranium-235, plutonium-239, and plutonium-241, and a relatively lower probability of neutron capture by uranium-238 (U-238) compared to the faster neutrons that originally result from fission, allowing use of low-enriched uranium or even natural uranium fuel. The moderator is often also the coolant, usually water under high pressure to increase the boiling point. These are surrounded by a reactor vessel, instrumentation to monitor and control the reactor, radiation shielding, and a containment building.

- Fast neutron reactors use fast neutrons to cause fission in their fuel. They do not have a neutron moderator, and use less-moderating coolants. Maintaining a chain reaction requires the fuel to be more highly enriched in fissile material (about 20% or more) due to the relatively lower probability of fission versus capture by U-238. Fast reactors have the potential to produce less transuranic waste because all actinides are fissionable with fast neutrons, but they are more difficult to build and more expensive to operate. Overall, fast reactors are less common than thermal reactors in most applications. Some early power stations were fast reactors, as are some Russian naval propulsion units. Construction of prototypes is continuing.

## Nuclear Fusion

Fusion power is an experimental technology, generally with hydrogen as fuel. While not suitable for power production, Farnsworth-Hirsch fusors are used to produce neutron radiation.

## Classification by Moderator Material

Used by thermal reactors:

- Graphite-moderated reactors
- Water moderated reactors
    - Heavy-water reactors (Used in Canada, India, Argentina, China, Pakistan, Romania and South Korea).)
    - Light-water-moderated reactors (LWRs). Light-water reactors (the most common type of thermal reactor) use ordinary water to moderate and cool the reactors. When at operating temperature, if the temperature of the water increases, its density drops, and fewer neutrons passing through it are slowed enough to trigger further reactions. That negative feedback stabilizes the reaction rate. Graphite and heavy-water reactors tend to be more thoroughly thermalized than light water reactors. Due to the extra thermalization, these types can use natural uranium/unenriched fuel.
- Light-element-moderated reactors.
    - Molten salt reactors (MSRs) are moderated by light elements such as lithium or beryllium, which are constituents of the coolant/fuel matrix salts LiF and $BeF_2$.
    - Liquid metal cooled reactors, such as those whose coolant is a mixture of lead and bismuth, may use BeO as a moderator.
- Organically moderated reactors (OMR) use biphenyl and terphenyl as moderator and coolant.

## Classification by Coolant

Treatment of the interior part of a VVER-1000 reactor frame on Atommash.

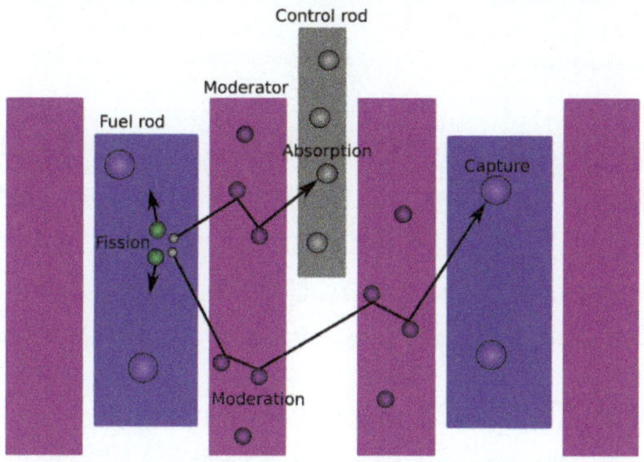


In thermal nuclear reactors (LWRs in specific), the coolant acts as a moderator that must slow down the neutrons before they can be efficiently absorbed by the fuel.

- Water cooled reactor. There are 104 operating reactors in the United States. Of these, 69 are pressurized water reactors (PWR), and 35 are boiling water reactors (BWR).
    - Pressurized water reactor (PWR) Pressurized water reactors constitute the large majority of all Western nuclear power plants.
        - A primary characteristic of PWRs is a pressurizer, a specialized pressure vessel. Most commercial PWRs and naval reactors use pressurizers. During normal operation, a pressurizer is partially filled with water, and a steam bubble is maintained above it by heating the water with submerged heaters. During normal operation, the pressurizer is connected to the primary reactor pressure vessel (RPV) and the pressurizer "bubble" provides an expansion space for changes in water volume in the reactor. This arrangement also provides a means of pressure control for the reactor by increasing or decreasing the steam pressure in the pressurizer using the pressurizer heaters.
        - Pressurised heavy water reactors are a subset of pressurized water reactors, sharing the use of a pressurized, isolated heat transport loop, but using heavy water as coolant and moderator for the greater neutron economies it offers.

- Boiling water reactor (BWR)
    - BWRs are characterized by boiling water around the fuel rods in the lower portion of a primary reactor pressure vessel. A boiling water reactor uses $^{235}U$, enriched as uranium dioxide, as its fuel. The fuel is assembled into rods housed in a steel vessel that is submerged in water. The nuclear fission causes the water to boil, generating steam. This steam flows through pipes into turbines. The turbines are driven by the steam, and this process generates electricity. During normal operation, pressure is controlled by the amount of steam flowing from the reactor pressure vessel to the turbine.
- Pool-type reactor

- Liquid metal cooled reactor. Since water is a moderator, it cannot be used as a coolant in a fast reactor. Liquid metal coolants have included sodium, NaK, lead, lead-bismuth eutectic, and in early reactors, mercury.
    - Sodium-cooled fast reactor
    - Lead-cooled fast reactor
- Gas cooled reactors are cooled by a circulating inert gas, often helium in high-temperature designs, while carbon dioxide has been used in past British and French nuclear power plants. Nitrogen has also been used. Utilization of the heat varies, depending on the reactor. Some reactors run hot enough that the gas can directly power a gas turbine. Older designs usually run the gas through a heat exchanger to make steam for a steam turbine.
- Molten salt reactors (MSRs) are cooled by circulating a molten salt, typically a eutectic mixture of fluoride salts, such as FLiBe. In a typical MSR, the coolant is also used as a matrix in which the fissile material is dissolved.

## Classification by Generation

- Generation I reactor (early prototypes, research reactors, non-commercial power producing reactors)
- Generation II reactor (most current nuclear power plants 1965–1996)
- Generation III reactor (evolutionary improvements of existing designs 1996-now)
- Generation IV reactor (technologies still under development unknown start date, possibly 2030)

In 2003, the French Commissariat à l'ÉnergieAtomique (CEA) was the first to refer to "Gen II" types in Nucleonics Week.

The first mentioning of "Gen III" was in 2000, in conjunction with the launch of the Generation IV International Forum (GIF) plans.

"Gen IV" was named in 2000, by the United States Department of Energy (DOE) for developing new plant types.

## Classification by Phase of Fuel

- Solid fueled
- Fluid fueled
  - Aqueous homogeneous reactor
  - Molten salt reactor
- Gas fueled (theoretical).

## Classification by Use

- Electricity
  - Nuclear power plants including small modular reactors
- Propulsion, see nuclear propulsion
  - Nuclear marine propulsion
  - Various proposed forms of rocket propulsion
- Other uses of heat
  - Desalination
  - Heat for domestic and industrial heating
  - Hydrogen production for use in a hydrogen economy
- Production reactors for transmutation of elements
  - Breeder reactors are capable of producing more fissile material than they consume during the fission chain reaction (by converting fertile U-238 to Pu-239, or Th-232 to U-233). Thus, a uranium breeder reactor, once running, can be re-fueled with natural or even depleted uranium, and a thorium breeder reactor can be re-fueled with thorium; however, an initial stock of fissile material is required.
  - Creating various radioactiveisotopes, such as americium for use in smoke detectors, and cobalt-60, molybdenum-99 and others, used for imaging and medical treatment.
  - Production of materials for nuclear weapons such as weapons-gradeplutonium
- Providing a source of neutron radiation (for example with the pulsed Godiva device) and positron radiation (e.g. neutron activation analysis and potassium-argon dating)
- Research reactor: Typically reactors used for research and training, materials testing, or the production of radioisotopes for medicine and industry. These are much smaller than power reactors or those propelling ships, and many are on university campuses. There are about 280 such reactors operating, in 56 countries. Some operate with high-enriched uranium fuel, and international efforts are underway to substitute low-enriched fuel.

## Current Technologies

Diablo Canyon — a PWR

Pressurized water reactors (PWR)

These reactors use a pressure vessel to contain the nuclear fuel, control rods, moderator, and coolant. They are cooled and moderated by high-pressure liquid water. The hot radioactive water that leaves the pressure vessel is looped through a steam generator, which in turn heats a secondary (non-radioactive) loop of water to steam that can run turbines. They are the majority of current reactors. This is a thermal neutron reactor design, the newest of which are the VVER-1200, Advanced Pressurized Water Reactor and the European Pressurized Reactor. United States Naval reactors are of this type.

Boiling water reactors (BWR)

A BWR is like a PWR without the steam generator. A boiling water reactor is cooled and moderated by water like a PWR, but at a lower pressure, which allows the water to boil inside the pressure vessel producing the steam that runs the turbines. Unlike a PWR, there is no primary and secondary loop. The thermal efficiency of these reactors can be higher, and they can be simpler, and even potentially more stable and safe. This is a thermal neutron reactor design, the newest of which are the Advanced Boiling Water Reactor and the Economic Simplified Boiling Water Reactor.

The CANDUQinshan Nuclear Power Plant

Pressurized Heavy Water Reactor (PHWR)

A Canadian design (known as CANDU), these reactors are heavy-water-cooled and -moderated pressurized-water reactors. Instead of using a single large pressure vessel as in a PWR, the fuel is contained in hundreds of pressure tubes. These reactors are fueled with natural uranium and are thermal neutron reactor designs. PHWRs can be refueled while at full power, which makes them very efficient in their use of uranium (it allows for precise flux control in the core). CANDU PHWRs have been built in Canada, Argentina, China, India, Pakistan, Romania, and South Korea. India also operates a number of PHWRs, often termed 'CANDU-derivatives', built after the Government of Canada halted nuclear dealings with India following the 1974 Smiling Buddha nuclear weapon test.

The Ignalina Nuclear Power Plant — a RBMK type (closed 2009)

ReaktorBolshoyMoschnostiKanalniy (High Power Channel Reactor) (RBMK)

A Soviet design, built to produce plutonium as well as power. RBMKs are water cooled with a graphite moderator. RBMKs are in some respects similar to CANDU in that they are refuelable during power operation and employ a pressure tube design instead of a PWR-style pressure vessel. However, unlike CANDU they are very unstable and large, making containment buildings for them expensive. A series of critical safety flaws have also been identified with the RBMK design, though some of these were corrected following the Chernobyl disaster. Their main attraction is their use of light water and un-enriched uranium. As of 2010, 11 remain open, mostly due to safety improvements and help from international safety agencies such as the DOE. Despite these safety improvements, RBMK reactors are still considered one of the most dangerous reactor designs in use. RBMK reactors were deployed only in the former Soviet Union.

The MagnoxSizewell A nuclear power station

The Torness nuclear power station — an AGR

Gas-cooled reactor (GCR) and advanced gas-cooled reactor (AGR)

These are generally graphite moderated and $CO_2$ cooled. They can have a high thermal efficiency compared with PWRs due to higher operating temperatures. There are a number of operating reactors of this design, mostly in the United Kingdom, where the concept was developed. Older designs (i.e. Magnox stations) are either shut down or will be in the near future. However, the AGCRs have an anticipated life of a further 10 to 20 years. This is a thermal neutron reactor design. Decommissioning costs can be high due to large volume of reactor core.

Liquid-metalfast-breeder reactor (LMFBR)

This is a reactor design that is cooled by liquid metal, totally unmoderated, and produces more fuel than it consumes. They are said to "breed" fuel, because they produce fissionable fuel during operation because of neutron capture. These reactors can function much like a PWR in terms of efficiency, and do not require much high-pressure containment, as the liquid metal does not need to be kept at high pressure, even at very high temperatures. BN-350 and BN-600 in USSR and Superphénix in France were a reactor of this type, as was Fermi-I in the United States. The Monju reactor in Japan suffered a sodium leak in 1995 and was restarted in May 2010. All of them use/used liquid sodium. These reactors are fast neutron, not thermal neutron designs. These reactors come in two types:

The Superphénix, one of the few FBRs

Lead-cooled

Using lead as the liquid metal provides excellent radiation shielding, and allows for operation at very high temperatures. Also, lead is (mostly) transparent to neutrons, so fewer neutrons are lost in the coolant, and the coolant does not become radioactive. Unlike sodium, lead is mostly inert, so there is less risk of explosion or accident, but such large quantities of lead may be problematic from toxicology and disposal points of view. Often a reactor of this type would use a lead-bismuth eutectic mixture. In this case, the bismuth would present some minor radiation problems, as it is not quite as transparent to neutrons, and can be transmuted to a radioactive isotope more readily than lead. The Russian Alfa class submarine uses a lead-bismuth-cooled fast reactor as its main power plant.

Sodium-cooled

Most LMFBRs are of this type. The sodium is relatively easy to obtain and work with, and it also manages to actually prevent corrosion on the various reactor parts immersed in it. However, sodium explodes violently when exposed to water, so care must be taken, but such explosions would not be vastly more violent than (for example) a leak of superheated fluid from a SCWR or PWR. EBR-I, the first reactor to have a core meltdown, was of this type.

Pebble-bed reactors (PBR)

These use fuel molded into ceramic balls, and then circulate gas through the balls. The result is an efficient, low-maintenance, very safe reactor with inexpensive, standardized fuel. The prototype was the AVR.

Molten salt reactors

These dissolve the fuels in fluoride salts, or use fluoride salts for coolant. These have many safety features, high efficiency and a high power density suitable for vehicles. Notably, they have no high pressures or flammable components in the core. The prototype was the MSRE, which also used the Thorium fuel cycle. As a breeder reactor type, it reprocesses the spent fuel, extracting both Uranium and transuranics, leaving only 0.1% of transuranic waste compared to conventional once-through uranium-fueled light water reactors currently in use. A separate issue are the radioactive fission products, which are not reprocessable and need to be disposed of as with conventional reactors.

Aqueous Homogeneous Reactor (AHR)

These reactors use soluble nuclear salts dissolved in water and mixed with a coolant and a neutron moderator.

## Future and Developing Technologies

## Advanced Reactors

More than a dozen advanced reactor designs are in various stages of development. Some are evolutionary from the PWR, BWR and PHWR designs above, some are more radical departures. The former include the advanced boiling water reactor (ABWR), two of which are now operating with others under construction, and the planned passively safeEconomic Simplified Boiling Water Reactor (ESBWR) and AP1000 units.

- The Integral Fast Reactor (IFR) was built, tested and evaluated during the 1980s and then retired under the Clinton administration in the 1990s due to nuclear non-proliferation policies of the administration. Recycling spent fuel is the core of its design and it therefore produces only a fraction of the waste of current reactors.
- The pebble-bed reactor, a high-temperature gas-cooled reactor (HTGCR), is designed so high temperatures reduce power output by Doppler broadening of the fuel's neutron cross-section. It uses ceramic fuels so its safe operating temperatures exceed the power-reduction temperature range. Most designs are cooled by inert helium. Helium is not subject to steam explosions, resists neutron absorption leading to radioactivity, and does not dissolve contaminants that can become radioactive. Typical designs have more layers (up to 7) of passive containment than light water reactors (usually 3). A unique feature that may aid safety is that the fuel-balls actually form the core's mechanism, and are replaced one-by-one as they age. The design of the fuel makes fuel reprocessing expensive.
- The Small, sealed, transportable, autonomous reactor (SSTAR) is being primarily researched and developed in the US, intended as a fast breeder reactor that is passively safe and could be remotely shut down in case the suspicion arises that it is being tampered with.
- The Clean And Environmentally Safe Advanced Reactor (CAESAR) is a nuclear reactor concept that uses steam as a moderator – this design is still in development.
- The Reduced moderation water reactor builds upon the Advanced boiling water reactor(ABWR) that is presently in use, it is not a complete fast reactor instead using mostly epithermal neutrons, which are between thermal and fast neutrons in speed.
- The hydrogen-moderated self-regulating nuclear power module (HPM) is a reactor design emanating from the Los Alamos National Laboratory that uses uranium hydride as fuel.
- Subcritical reactors are designed to be safer and more stable, but pose a number of engineering and economic difficulties. One example is the Energy amplifier.
- Thorium-based reactors. It is possible to convert Thorium-232 into U-233 in reactors specially designed for the purpose. In this way, thorium, which is four times more abundant than uranium, can be used to breed U-233 nuclear fuel. U-233 is also believed to have favourable nuclear properties as compared to traditionally used U-235, including better neutron economy and lower production of long lived transuranic waste.
    - Advanced heavy-water reactor (AHWR)— A proposed heavy water moderated nuclear power reactor that will be the next generation design of the PHWR type. Under development in the Bhabha Atomic Research Centre (BARC), India.
    - KAMINI — A unique reactor using Uranium-233 isotope for fuel. Built in India by BARC and Indira Gandhi Center for Atomic Research (IGCAR).
    - India is also planning to build fast breeder reactors using the thorium – Uranium-233 fuel cycle. The FBTR (Fast Breeder Test Reactor) in operation at Kalpakkam (India) uses Plutonium as a fuel and liquid sodium as a coolant.

## Generation IV Reactors

Generation IV reactors are a set of theoretical nuclear reactor designs currently being researched. These designs are generally not expected to be available for commercial construction before 2030. Current reactors in operation around the world are generally considered second- or third-generation systems, with the first-generation systems having been retired some time ago. Research into these reactor types was officially started by the Generation IV International Forum (GIF) based on eight technology goals. The primary goals being to improve nuclear safety, improve proliferation resistance, minimize waste and natural resource utilization, and to decrease the cost to build and run such plants.

- Gas-cooled fast reactor
- Lead-cooled fast reactor
- Molten salt reactor
- Sodium-cooled fast reactor
- Supercritical water reactor
- Very-high-temperature reactor.

## Generation V+ Reactors

Generation V reactors are designs which are theoretically possible, but which are not being actively considered or researched at present. Though such reactors could be built with current or near term technology, they trigger little interest for reasons of economics, practicality, or safety.

- Liquid-core reactor. A closed loop liquid-core nuclear reactor, where the fissile material is molten uranium or uranium solution cooled by a working gas pumped in through holes in the base of the containment vessel.
- Gas-core reactor. A closed loop version of the nuclear lightbulb rocket, where the fissile material is gaseous uranium-hexafluoride contained in a fused silica vessel. A working gas (such as hydrogen) would flow around this vessel and absorb the UV light produced by the reaction. This reactor design could also function as a rocket engine, as featured in Harry Harrison's 1976 science-fiction novel 'Skyfall'. In theory, using $UF_6$ as a working fuel directly (rather than as a stage to one, as is done now) would mean lower processing costs, and very small reactors. In practice, running a reactor at such high power densities would probably produce unmanageable neutron flux, weakening most reactor materials, and therefore as the flux would be similar to that expected in fusion reactors, it would require similar materials to those selected by the International Fusion Materials Irradiation Facility.
    - Gas core EM reactor. As in the gas core reactor, but with photovoltaic arrays converting the UV light directly to electricity.
- Fission fragment reactor.
- Hybrid nuclear fusion. Would use the neutrons emitted by fusion to fission a blanket of fertile material, like U-238 or Th-232 and transmutate other reactor's spent nuclear fuel/ nuclear waste into relatively more benign isotopes.

## Fusion Reactors

Controlled nuclear fusion could in principle be used in fusion power plants to produce power without the complexities of handling actinides, but significant scientific and technical obstacles remain. Several fusion reactors have been built, but only recently reactors have been able to release more energy than the amount of energy used in the process. Despite research having started in the 1950s, no commercial fusion reactor is expected before 2050. The ITER project is currently leading the effort to harness fusion power.

## Nuclear Fuel Cycle

Thermal reactors generally depend on refined and enriched uranium. Some nuclear reactors can operate with a mixture of plutonium and uranium. The process by which uranium ore is mined, processed, enriched, used, possibly reprocessed and disposed of is known as the nuclear fuel cycle.

Under 1% of the uranium found in nature is the easily fissionable U-235 isotope and as a result most reactor designs require enriched fuel. Enrichment involves increasing the percentage of U-235 and is usually done by means of gaseous diffusion or gas centrifuge. The enriched result is then converted into uranium dioxide powder, which is pressed and fired into pellet form. These pellets are stacked into tubes which are then sealed and called fuel rods. Many of these fuel rods are used in each nuclear reactor.

Most BWR and PWR commercial reactors use uranium enriched to about 4% U-235, and some commercial reactors with a high neutron economy do not require the fuel to be enriched at all (that is, they can use natural uranium). According to the International Atomic Energy Agency there are at least 100 research reactors in the world fueled by highly enriched (weapons-grade/90% enrichment uranium). Theft risk of this fuel (potentially used in the production of a nuclear weapon) has led to campaigns advocating conversion of this type of reactor to low-enrichment uranium (which poses less threat of proliferation).

Fissile U-235 and non-fissile but fissionable and fertile U-238 are both used in the fission process. U-235 is fissionable by thermal (i.e. slow-moving) neutrons. A thermal neutron is one which is moving about the same speed as the atoms around it. Since all atoms vibrate proportionally to their absolute temperature, a thermal neutron has the best opportunity to fission U-235 when it is moving at this same vibrational speed. On the other hand, U-238 is more likely to capture a neutron when the neutron is moving very fast. This U-239 atom will soon decay into plutonium-239, which is another fuel. Pu-239 is a viable fuel and must be accounted for even when a highly enriched uranium fuel is used. Plutonium fissions will dominate the U-235 fissions in some reactors, especially after the initial loading of U-235 is spent. Plutonium is fissionable with both fast and thermal neutrons, which make it ideal for either nuclear reactors or nuclear bombs.

Most reactor designs in existence are thermal reactors and typically use water as a neutron moderator (moderator means that it slows down the neutron to a thermal speed) and as a coolant. But in a fast breeder reactor, some other kind of coolant is used which will not moderate or slow the neutrons down much. This enables fast neutrons to dominate, which can effectively be used to constantly replenish the fuel supply. By merely placing cheap unenriched uranium into such a core, the non-fissionable U-238 will be turned into Pu-239, "breeding" fuel.

In thorium fuel cyclethorium-232 absorbs a neutron in either a fast or thermal reactor. The thorium-233 beta decays to protactinium-233 and then to uranium-233, which in turn is used as fuel. Hence, like uranium-238, thorium-232 is a fertile material.

## Fueling of Nuclear Reactors

The amount of energy in the reservoir of nuclear fuel is frequently expressed in terms of "full-power days," which is the number of 24-hour periods (days) a reactor is scheduled for operation at full power output for the generation of heat energy. The number of full-power days in a reactor's operating cycle (between refueling outage times) is related to the amount of fissileuranium-235 (U-235) contained in the fuel assemblies at the beginning of the cycle. A higher percentage of U-235 in the core at the beginning of a cycle will permit the reactor to be run for a greater number of full-power days.

At the end of the operating cycle, the fuel in some of the assemblies is "spent" and is discharged and replaced with new (fresh) fuel assemblies, although in practice it is the buildup of reaction poisons in nuclear fuel that determines the lifetime of nuclear fuel in a reactor. Long before all possible fission has taken place, the buildup of long-lived neutron absorbing fission byproducts impedes the chain reaction. The fraction of the reactor's fuel core replaced during refueling is typically one-fourth for a boiling-water reactor and one-third for a pressurized-water reactor. The disposition and storage of this spent fuel is one of the most challenging aspects of the operation of a commercial nuclear power plant. This nuclear waste is highly radioactive and its toxicity presents a danger for thousands of years.

Not all reactors need to be shut down for refueling; for example, pebble bed reactors, RBMK reactors, molten salt reactors, Magnox, AGR and CANDU reactors allow fuel to be shifted through the reactor while it is running. In a CANDU reactor, this also allows individual fuel elements to be situated within the reactor core that are best suited to the amount of U-235 in the fuel element.

The amount of energy extracted from nuclear fuel is called its burnup, which is expressed in terms of the heat energy produced per initial unit of fuel weight. Burn up is commonly expressed as megawatt days thermal per metric ton of initial heavy metal.

## Nuclear Safety Concerns and Controversy

Nuclear safety covers the actions taken to prevent nuclear and radiation accidents or to limit their consequences. The nuclear power industry has improved the safety and performance of reactors, and has proposed new safer (but generally untested) reactor designs but there is no guarantee that the reactors will be designed, built and operated correctly. Mistakes do occur and the designers of reactors at Fukushima in Japan did not anticipate that a tsunami generated by an earthquake would disable the backup systems that were supposed to stabilize the reactor after the earthquake, despite multiple warnings by the NRG and the Japanese nuclear safety administration. According to UBS AG, the Fukushima I nuclear accidents have cast doubt on whether even an advanced economy like Japan can master nuclear safety. Catastrophic scenarios involving terrorist attacks are also conceivable. An interdisciplinary team from MIT has estimated that given the expected growth of nuclear power from 2005–2055, at least four serious nuclear accidents would be expected in that period.

## Nuclear Accidents and Controversy

Three of the reactors at Fukushima I overheated, causing meltdowns that eventually led to explosions, which released large amounts of radioactive material into the air.

Some serious nuclear and radiation accidents have occurred. Nuclear power plant accidents include the SL-1 accident (1961), the Three Mile Island accident (1979), Chernobyl disaster (1986), and the Fukushima Daiichi nuclear disaster (2011).Nuclear-powered submarine mishaps include the K-19 reactor accident (1961), the K-27 reactor accident (1968), and the K-431 reactor accident (1985).

Nuclear reactors have been launched into Earth orbit at least 34 times. A number of incidents connected with the unmanned nuclear-reactor-powered Soviet RORSAT radar satellite program resulted in spent nuclear fuel re-entering the Earth's atmosphere from orbit.

## Natural Nuclear Reactors

Although nuclear fission reactors are often thought of as being solely a product of modern technology, the first nuclear fission reactors were in fact naturally occurring. A natural nuclear fission reactor can occur under certain circumstances that mimic the conditions in a constructed reactor. Fifteen natural fission reactors have so far been found in three separate ore deposits at the Oklo uranium mine in Gabon, West Africa. First discovered in 1972 by French physicist Francis Perrin, they are collectively known as the Oklo Fossil Reactors. Self-sustaining nuclear fission reactions took place in these reactors approximately 1.5 billion years ago, and ran for a few hundred thousand years, averaging 100 kW of power output during that time. The concept of a natural nuclear reactor was theorized as early as 1956 by Paul Kuroda at the University of Arkansas.

Such reactors can no longer form on Earth: radioactive decay over this immense time span has reduced the proportion of U-235 in naturally occurring uranium to below the amount required to sustain a chain reaction.

The natural nuclear reactors formed when a uranium-rich mineral deposit became inundated with groundwater that acted as a neutron moderator, and a strong chain reaction took place. The water moderator would boil away as the reaction increased, slowing it back down again and preventing a meltdown. The fission reaction was sustained for hundreds of thousands of years.

These natural reactors are extensively studied by scientists interested in geologic radioactive waste disposal. They offer a case study of how radioactive isotopes migrate through the Earth's crust.

This is a significant area of controversy as opponents of geologic waste disposal fear that isotopes from stored waste could end up in water supplies or be carried into the environment.

## Emission

Nuclear reactor emits tiny amount of tritium, Sr-90 to air and groundwater.Water contaminated with tritium is colorless and odorless. Large doses of Sr-90increases the risk of bone cancer and leukemia in animals, and is presumed to do so in people.

## Nuclear Fission

In nuclear physics and nuclear chemistry, nuclear fission is either a nuclear reaction or a radioactive decay process in which the nucleus of an atom splits into smaller parts (lighter nuclei). The fission process often produces free neutrons and gammaphotons, and releases a very large amount of energy even by the energetic standards of radioactive decay.

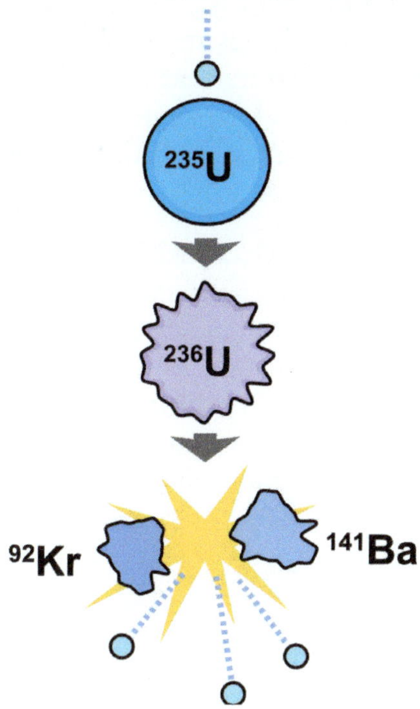


An induced fission reaction. A neutron is absorbed by a uranium-235 nucleus, turning it briefly into an excited uranium-236 nucleus, with the excitation energy provided by the kinetic energy of the neutron plus the forces that bind the neutron. The uranium-236, in turn, splits into fast-moving lighter elements (fission products) and releases three free neutrons. At the same time, one or more "prompt gamma rays" (not shown) are produced, as well.

Nuclear fission of heavy elements was discovered on December 17, 1938 by German Otto Hahn and his assistant Fritz Strassmann, and explained theoretically in January 1939 by Lise Meitner and her nephew Otto Robert Frisch. Frisch named the process by analogy with biological fission of living cells. It is an exothermic reaction which can release large amounts of energy both as electromagnetic radiation and as kinetic energy of the fragments (heating the bulk material where fission takes place). In order for fission to produce energy, the total binding energy of the resulting elements must be less negative (higher energy) than that of the starting element.

Fission is a form of nuclear transmutation because the resulting fragments are not the same element as the original atom. The two nuclei produced are most often of comparable but slightly different sizes, typically with a mass ratio of products of about 3 to 2, for common fissileisotopes. Most fissions are binary fissions (producing two charged fragments), but occasionally (2 to 4 times per 1000 events), *three* positively charged fragments are produced, in a ternary fission. The smallest of these fragments in ternary processes ranges in size from a proton to an argon nucleus.

Apart from fission induced by a neutron, harnessed and exploited by humans, a natural form of spontaneous radioactive decay (not requiring a neutron) is also referred to as fission, and occurs especially in very high-mass-number isotopes. Spontaneous fission was discovered in 1940 by Flyorov, Petrzhak and Kurchatov in Moscow, when they decided to confirm that, without bombardment by neutrons, the fission rate of uranium was indeed negligible, as predicted by Niels Bohr; it wasn't.

The unpredictable composition of the products (which vary in a broad probabilistic and somewhat chaotic manner) distinguishes fission from purely quantum-tunneling processes such as proton emission, alpha decay, and cluster decay, which give the same products each time. Nuclear fission produces energy for nuclear power and drives the explosion of nuclear weapons. Both uses are possible because certain substances called nuclear fuels undergo fission when struck by fission neutrons, and in turn emit neutrons when they break apart. This makes possible a self-sustaining nuclear chain reaction that releases energy at a controlled rate in a nuclear reactor or at a very rapid uncontrolled rate in a nuclear weapon.

The amount of free energy contained in nuclear fuel is millions of times the amount of free energy contained in a similar mass of chemical fuel such as gasoline, making nuclear fission a very dense source of energy. The products of nuclear fission, however, are on average far more radioactive than the heavy elements which are normally fissioned as fuel, and remain so for significant amounts of time, giving rise to a nuclear waste problem. Concerns over nuclear waste accumulation and over the destructive potential of nuclear weapons are a counterbalance to the peaceful desire to use fission as an energy source, and give rise to ongoing politicaldebate over nuclear power.

## Physical Overview

### Mechanism

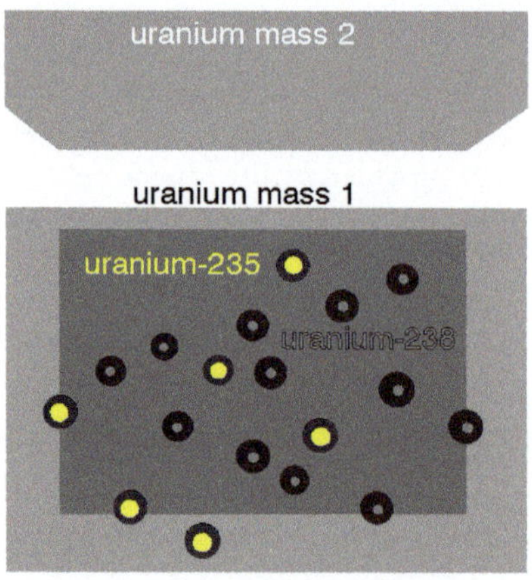


A visual representation of an induced nuclear fission event where a slow-moving neutron is absorbed by the nucleus of a uranium-235 atom, which fissions into two fast-moving lighter elements (fission products) and additional neutrons. Most of the energy released is in the form of the kinetic velocities of the fission products and the neutrons.

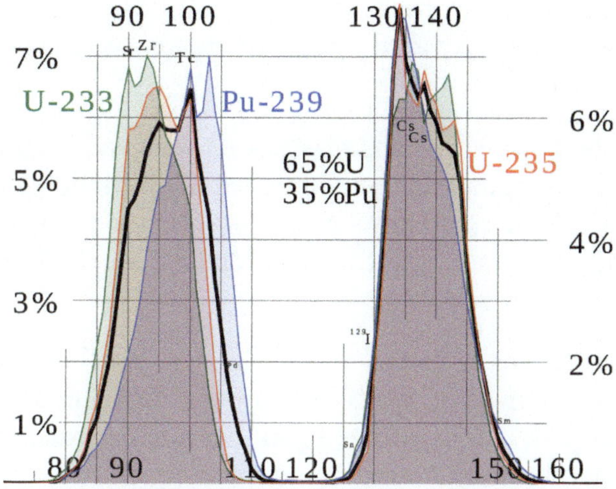


Fission product yields by mass for thermal neutron fission of U-235, Pu-239, a combination of the two typical of current nuclear power reactors, and U-233 used in the thorium cycle.

Nuclear fission can occur without neutron bombardment as a type of radioactive decay. This type of fission (called spontaneous fission) is rare except in a few heavy isotopes. In engineered nuclear devices, essentially all nuclear fission occurs as a "nuclear reaction" — a bombardment-driven process that results from the collision of two subatomic particles. In nuclear reactions, a subatomic particle collides with an atomic nucleus and causes changes to it. Nuclear reactions are thus driven by the mechanics of bombardment, not by the relatively constant exponential decay and half-life characteristic of spontaneous radioactive processes.

Many types of nuclear reactions are currently known. Nuclear fission differs importantly from other types of nuclear reactions, in that it can be amplified and sometimes controlled via a nuclear chain reaction (one type of general chain reaction). In such a reaction, free neutrons released by each fission event can trigger yet more events, which in turn release more neutrons and cause more fissions.

The chemical elementisotopes that can sustain a fission chain reaction are called nuclear fuels, and are said to be *fissile*. The most common nuclear fuels are $^{235}U$ (the isotope of uranium with an atomic mass of 235 and of use in nuclear reactors) and $^{239}Pu$ (the isotope of plutonium with an atomic mass of 239). These fuels break apart into a bimodal range of chemical elements with atomic masses centering near 95 and 135 u (fission products). Most nuclear fuels undergo spontaneous fission only very slowly, decaying instead mainly via an alpha-betadecay chain over periods of millennia to eons. In a nuclear reactor or nuclear weapon, the overwhelming majority of fission events are induced by bombardment with another particle, a neutron, which is itself produced by prior fission events.

Nuclear fissions in fissile fuels are the result of the nuclear excitation energy produced when a fissile nucleus captures a neutron. This energy, resulting from the neutron capture, is a result of the attractive nuclear force acting between the neutron and nucleus. It is enough to deform the nucleus into a double-lobed "drop," to the point that nuclear fragments exceed the distances at which the nuclear force can hold two groups of charged nucleons together and, when this happens, the two fragments complete their separation and then are driven further apart by their mutually repulsive charges, in a process which becomes irreversible with greater and greater distance. A similar process occurs in fissionable isotopes (such as uranium-238), but in order to fission, these isotopes require additional energy provided by fast neutrons (such as those produced by nuclear fusion in thermonuclear weapons).

The liquid drop model of the atomic nucleus predicts equal-sized fission products as an outcome of nuclear deformation. The more sophisticated nuclear shell model is needed to mechanistically explain the route to the more energetically favorable outcome, in which one fission product is slightly smaller than the other. A theory of the fission based on shell model has been formulated by Maria Goeppert Mayer.

The most common fission process is binary fission, and it produces the fission products noted above, at 95±15 and 135±15 u. However, the binary process happens merely because it is the most probable. In anywhere from 2 to 4 fissions per 1000 in a nuclear reactor, a process called ternary fission produces three positively charged fragments (plus neutrons) and the smallest of these may range from so small a charge and mass as a proton (Z=1), to as large a fragment as argon (Z=18). The most common small fragments, however, are composed of 90% helium-4 nuclei with more energy than alpha particles from alpha decay (so-called "long range alphas" at ~ 16 MeV), plus helium-6 nuclei, and tritons (the nuclei of tritium). The ternary process is less common, but still ends up producing significant helium-4 and tritium gas buildup in the fuel rods of modern nuclear reactors.

## Energetics

### Input

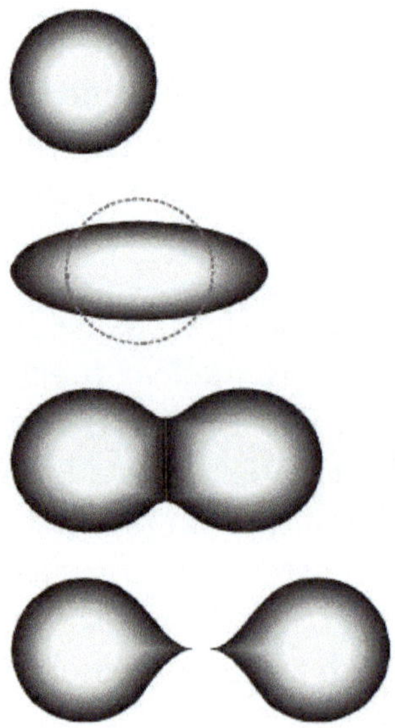

The stages of binary fission in a liquid drop model. Energy input deforms the nucleus into a fat "cigar" shape, then a "peanut" shape, followed by binary fission as the two lobes exceed the short-range nuclear force attraction distance, then are pushed apart and away by their electrical charge. In the liquid drop model, the two fission fragments are predicted to be the same size. The nuclear shell model allows for them to differ in size, as usually experimentally observed.

The fission of a heavy nucleus requires a total input energy of about 7 to 8 million electron volts (MeV) to initially overcome the nuclear force which holds the nucleus into a spherical or nearly spherical shape, and from there, deform it into a two-lobed ("peanut") shape in which the lobes are able to continue to separate from each other, pushed by their mutual positive charge, in the most common process of binary fission (two positively charged fission products + neutrons). Once the nuclear lobes have been pushed to a critical distance, beyond which the short range strong force can no longer hold them together, the process of their separation proceeds from the energy of the (longer range) electromagnetic repulsion between the fragments. The result is two fission fragments moving away from each other, at high energy.

About 6 MeV of the fission-input energy is supplied by the simple binding of an extra neutron to the heavy nucleus via the strong force; however, in many fissionable isotopes, this amount of energy is not enough for fission. Uranium-238, for example, has a near-zero fission cross section for neutrons of less than one MeV energy. If no additional energy is supplied by any other mechanism, the nucleus will not fission, but will merely absorb the neutron, as happens when U-238 absorbs slow and even some fraction of fast neutrons, to become U-239. The remaining energy to initiate fission can be supplied by two other mechanisms: one of these is more kinetic energy of the incoming neutron, which is increasingly able to fission a fissionable heavy nucleus as it exceeds a kinetic energy of one MeV or more (so-called fast neutrons). Such high energy neutrons are able to fission U-238 directly. However, this process cannot happen to a great extent in a nuclear reactor, as too small a fraction of the fission neutrons produced by any type of fission have enough energy to efficiently fission U-238 (fission neutrons have a mode energy of 2 MeV, but a median of only 0.75 MeV, meaning half of them have less than this insufficient energy).

Among the heavy actinide elements, however, those isotopes that have an odd number of neutrons (such as U-235 with 143 neutrons) bind an extra neutron with an additional 1 to 2 MeV of energy over an isotope of the same element with an even number of neutrons (such as U-238 with 146 neutrons). This extra binding energy is made available as a result of the mechanism of neutron pairing effects. This extra energy results from the Pauli exclusion principle allowing an extra neutron to occupy the same nuclear orbital as the last neutron in the nucleus, so that the two form a pair. In such isotopes, therefore, no neutron kinetic energy is needed, for all the necessary energy is supplied by absorption of any neutron, either of the slow or fast variety (the former are used in moderated nuclear reactors, and the latter are used in fast neutron reactors, and in weapons). As noted above, the subgroup of fissionable elements that may be fissioned efficiently with their own fission neutrons (thus potentially causing a nuclear chain reaction in relatively small amounts of the pure material) are termed "fissile." Examples of fissile isotopes are U-235 and plutonium-239.

## Output

Typical fission events release about two hundred million eV (200 MeV) of energy for each fission event. The exact isotope which is fissioned, and whether or not it is fissionable or fissile, has only a small impact on the amount of energy released. This can be easily seen by examining the curve of binding energy, and noting that the average binding energy of the actinide nuclides beginning with uranium is around 7.6 MeV per nucleon. Looking further left on the curve of binding energy, where the fission products cluster, it is easily observed that the binding energy of the fission products tends to center around 8.5 MeV per nucleon. Thus, in any fission event of an isotope in the actinide's range of mass, roughly 0.9 MeV is released per nucleon of the starting element. The fission of U235 by a slow neutron yields nearly identical energy to the fission of U238 by a fast neutron. This energy release profile holds true for thorium and the various minor actinides as well.

By contrast, most chemicaloxidation reactions (such as burning coal or TNT) release at most a few eV per event. So, nuclear fuel contains at least ten million times more usable energy per unit mass than does chemical fuel. The energy of nuclear fission is released as kinetic energy of the fission products and fragments, and as electromagnetic radiation in the form of gamma rays; in a nuclear reactor, the energy is converted to heat as the particles and gamma rays collide with the atoms that make up the reactor and its working fluid, usually water or occasionally heavy water or molten salts.

When a uranium nucleus fissions into two daughter nuclei fragments, about 0.1 percent of the mass of the uranium nucleus appears as the fission energy of ~200 MeV. For uranium-235 (total mean fission energy 202.5 MeV), typically ~169 MeV appears as the kinetic energy of the daughter nuclei, which fly apart at about 3% of the speed of light, due to Coulomb repulsion. Also, an average of 2.5 neutrons are emitted, with a mean kinetic energy per neutron of ~2 MeV (total of 4.8 MeV). The fission reaction also releases ~7 MeV in prompt gamma rayphotons. The latter figure means that a nuclear fission explosion or criticality accident emits about 3.5% of its energy as gamma rays, less than 2.5% of its energy as fast neutrons (total of both types of radiation ~ 6%), and the rest as kinetic energy of fission fragments (this appears almost immediately when the fragments impact surrounding matter, as simple heat). In an atomic bomb, this heat may serve to raise the temperature of the bomb core to 100 million kelvin and cause secondary emission of soft X-rays, which convert some of this energy to ionizing radiation. However, in nuclear reactors, the fission fragment kinetic energy remains as low-temperature heat, which itself causes little or no ionization.

So-called neutron bombs (enhanced radiation weapons) have been constructed which release a larger fraction of their energy as ionizing radiation (specifically, neutrons), but these are all thermonuclear devices which rely on the nuclear fusion stage to produce the extra radiation. The energy dynamics of pure fission bombs always remain at about 6% yield of the total in radiation, as a prompt result of fission.

The total *prompt fission* energy amounts to about 181 MeV, or ~ 89% of the total energy which is eventually released by fission over time. The remaining ~ 11% is released in beta decays which have various half-lives, but begin as a process in the fission products immediately; and in delayed gamma emissions associated with these beta decays. For example, in uranium-235 this delayed energy is divided into about 6.5 MeV in betas, 8.8 MeV in antineutrinos (released at the same time as the betas), and finally, an additional 6.3 MeV in delayed gamma emission from the excited beta-decay products (for a mean total of ~10 gamma ray emissions per fission, in all). Thus, about 6.5% of the total energy of fission is released some time after the event, as non-prompt or delayed ionizing radiation, and the delayed ionizing energy is about evenly divided between gamma and beta ray energy.

In a reactor that has been operating for some time, the radioactive fission products will have built up to steady state concentrations such that their rate of decay is equal to their rate of formation, so that their fractional total contribution to reactor heat (via beta decay) is the same as these radioisotopic fractional contributions to the energy of fission. Under these conditions, the 6.5% of fission which appears as delayed ionizing radiation (delayed gammas and betas from radioactive fission products) contributes to the steady-state reactor heat production under power. It is this output fraction which remains when the reactor is suddenly shut down (undergoes scram). For this reason, the reactor decay heat output begins at 6.5% of the full reactor steady state fission power, once the reactor is shut down. However, within hours, due to decay of these isotopes, the decay power output is far less.

The remainder of the delayed energy (8.8 MeV/202.5 MeV = 4.3% of total fission energy) is emitted as antineutrinos, which as a practical matter, are not considered "ionizing radiation." The reason is that energy released as antineutrinos is not captured by the reactor material as heat, and escapes directly through all materials (including the Earth) at nearly the speed of light, and into interplanetary space (the amount absorbed is minuscule). Neutrino radiation is ordinarily not classed as ionizing radiation, because it is almost entirely not absorbed and therefore does not produce effects (although the very rare neutrino event is ionizing). Almost all of the rest of the radiation (6.5% delayed beta and gamma radiation) is eventually converted to heat in a reactor core or its shielding.

Some processes involving neutrons are notable for absorbing or finally yielding energy — for example neutron kinetic energy does not yield heat immediately if the neutron is captured by a uranium-238 atom to breed plutonium-239, but this energy is emitted if the plutonium-239 is later fissioned. On the other hand, so-called delayed neutrons emitted as radioactive decay products with half-lives up to several minutes, from fission-daughters, are very important to reactor control, because they give a characteristic "reaction" time for the total nuclear reaction to double in size, if the reaction is run in a "delayed-critical" zone which deliberately relies on these neutrons for a supercritical chain-reaction (one in which each fission cycle yields more neutrons than it absorbs). Without their existence, the nuclear chain-reaction would be prompt critical and increase in size faster than it could be controlled by human intervention. In this case, the first experimental atomic reactors would have run away to a dangerous and messy "prompt critical reaction" before their operators could have manually shut them down (for this reason, designer Enrico Fermi included radiation-counter-triggered control rods, suspended by electromagnets, which could automatically drop into the center of Chicago Pile-1). If these delayed neutrons are captured without producing fissions, they produce heat as well.

## Product Nuclei and Binding Energy

In fission there is a preference to yield fragments with even proton numbers, which is called the odd-even effect on the fragments' charge distribution. However, no odd-even effect is observed on fragment mass number distribution. This result is attributed to nucleon pair breaking.

In nuclear fission events the nuclei may break into any combination of lighter nuclei, but the most common event is not fission to equal mass nuclei of about mass 120; the most common event (depending on isotope and process) is a slightly unequal fission in which one daughter nucleus has a mass of about 90 to 100 u and the other the remaining 130 to 140 u. Unequal fissions are energetically more favorable because this allows one product to be closer to the energetic minimum near mass 60 u (only a quarter of the average fissionable mass), while the other nucleus with mass 135 u is still not far out of the range of the most tightly bound nuclei (another statement of this, is that the atomic binding energy curve is slightly steeper to the left of mass 120 u than to the right of it).

## Origin of the Active Energy and the Curve of Binding Energy

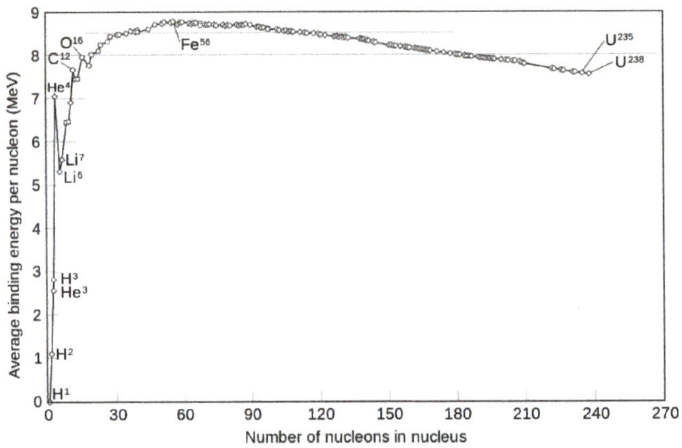

The "curve of binding energy": A graph of binding energy per nucleon of common isotopes.

Nuclear fission of heavy elements produces energy because the specific binding energy (binding energy per mass) of intermediate-mass nuclei with atomic numbers and atomic masses close to $^{62}$Ni and $^{56}$Fe is greater than the nucleon-specific binding energy of very heavy nuclei, so that energy is released when heavy nuclei are broken apart. The total rest masses of the fission products (Mp) from a single reaction is less than the mass of the original fuel nucleus (M). The excess mass $\Delta m = M - Mp$ is the invariant mass of the energy that is released as photons (gamma rays) and kinetic energy of the fission fragments, according to the mass-energy equivalence formula $E = mc^2$.

The variation in specific binding energy with atomic number is due to the interplay of the two fundamental forces acting on the component nucleons (protons and neutrons) that make up the nucleus. Nuclei are bound by an attractive nuclear force between nucleons, which overcomes the electrostatic repulsion between protons. However, the nuclear force acts only over relatively short ranges (a few nucleon diameters), since it follows an exponentially decaying Yukawa potential which makes it insignificant at longer distances. The electrostatic repulsion is of longer range, since it decays by an inverse-square rule, so that nuclei larger than about 12 nucleons in diameter reach a point that the total electrostatic repulsion overcomes the nuclear force and causes them to be spontaneously unstable. For the same reason, larger nuclei (more than about eight nucleons in diameter) are less tightly bound per unit mass than are smaller nuclei; breaking a large nucleus into two or more intermediate-sized nuclei releases energy. The origin of this energy is the nuclear force, which intermediate-sized nuclei allows to act more efficiently, because each nucleon has more neighbors which are within the short range attraction of this force. Thus less energy is needed in the smaller nuclei and the difference to the state before is set free.

Also because of the short range of the strong binding force, large stable nuclei must contain proportionally more neutrons than do the lightest elements, which are most stable with a 1 to 1 ratio of protons and neutrons. Nuclei which have more than 20 protons cannot be stable unless they have more than an equal number of neutrons. Extra neutrons stabilize heavy elements because they add to strong-force binding (which acts between all nucleons) without adding to proton–proton repulsion. Fission products have, on average, about the same ratio of neutrons and protons as their parent nucleus, and are therefore usually unstable to beta decay (which changes neutrons to protons) because they have proportionally too many neutrons compared to stable isotopes of similar mass.

This tendency for fission product nuclei to beta-decay is the fundamental cause of the problem of radioactivehigh level waste from nuclear reactors. Fission products tend to be beta emitters, emitting fast-moving electrons to conserve electric charge, as excess neutrons convert to protons in the fission-product atoms.

## Chain Reactions

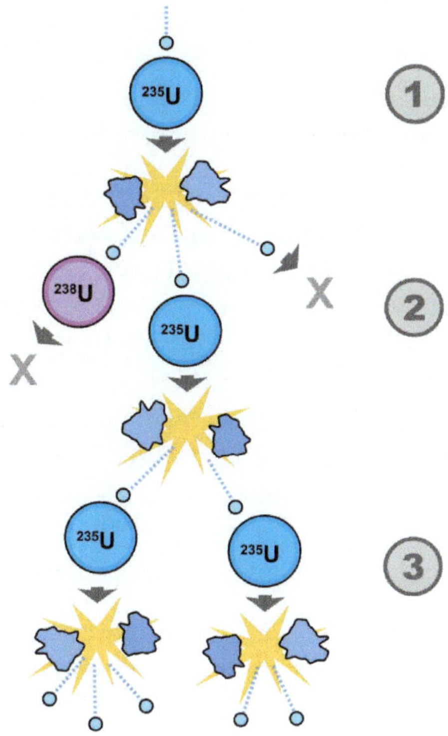


A schematic nuclear fission chain reaction. 1. A uranium-235 atom absorbs a neutron and fissions into two new atoms (fission fragments), releasing three new neutrons and some binding energy. 2. One of those neutrons is absorbed by an atom of uranium-238 and does not continue the reaction. Another neutron is simply lost and does not collide with anything, also not continuing the reaction. However, the one neutron does collide with an atom of uranium-235, which then fissions and releases two neutrons and some binding energy. 3. Both of those neutrons collide with uranium-235 atoms, each of which fissions and releases between one and three neutrons, which can then continue the reaction.

Several heavy elements, such as uranium, thorium, and plutonium, undergo both spontaneous fission, a form of radioactive decay and *induced fission*, a form of nuclear reaction. Elemental isotopes that undergo induced fission when struck by a free neutron are called fissionable; isotopes that undergo fission when struck by a slow-moving thermal neutron are also called fissile. A few particularly fissile and readily obtainable isotopes (notably $^{233}U$, $^{235}U$ and $^{239}Pu$) are called nuclear fuels because they can sustain a chain reaction and can be obtained in large enough quantities to be useful.

All fissionable and fissile isotopes undergo a small amount of spontaneous fission which releases a few free neutrons into any sample of nuclear fuel. Such neutrons would escape rapidly from the fuel and become a free neutron, with a mean lifetime of about 15 minutes before decaying to protons and beta particles. However, neutrons almost invariably impact and are absorbed by other nuclei in the vicinity long before this happens (newly created fission neutrons move at about 7% of the speed of light, and even moderated neutrons move at about 8 times the speed of sound). Some neutrons will impact fuel nuclei and induce further fissions, releasing yet more neutrons. If enough nuclear fuel is assembled in one place, or if the escaping neutrons are sufficiently contained, then these freshly emitted neutrons outnumber the neutrons that escape from the assembly, and a *sustained nuclear chain reaction* will take place.

An assembly that supports a sustained nuclear chain reaction is called a critical assembly or, if the assembly is almost entirely made of a nuclear fuel, a critical mass. The word "critical" refers to a cusp in the behavior of the differential equation that governs the number of free neutrons present in the fuel: if less than a critical mass is present, then the amount of neutrons is determined by radioactive decay, but if a critical mass or more is present, then the amount of neutrons is controlled instead by the physics of the chain reaction. The actual mass of a *critical mass* of nuclear fuel depends strongly on the geometry and surrounding materials.

Not all fissionable isotopes can sustain a chain reaction. For example, $^{238}U$, the most abundant form of uranium, is fissionable but not fissile: it undergoes induced fission when impacted by an energetic neutron with over 1 MeV of kinetic energy. However, too few of the neutrons produced by $^{238}U$ fission are energetic enough to induce further fissions in $^{238}U$, so no chain reaction is possible with this isotope. Instead, bombarding $^{238}U$ with slow neutrons causes it to absorb them (becoming $^{239}U$) and decay by beta emission to $^{239}Np$ which then decays again by the same process to $^{239}Pu$; that process is used to manufacture $^{239}Pu$ in breeder reactors. In-situ plutonium production also contributes to the neutron chain reaction in other types of reactors after sufficient plutonium-239 has been produced, since plutonium-239 is also a fissile element which serves as fuel. It is estimated that up to half of the power produced by a standard "non-breeder" reactor is produced by the fission of plutonium-239 produced in place, over the total life-cycle of a fuel load.

Fissionable, non-fissile isotopes can be used as fission energy source even without a chain reaction. Bombarding $^{238}U$ with fast neutrons induces fissions, releasing energy as long as the external neutron source is present. This is an important effect in all reactors where fast neutrons from the fissile isotope can cause the fission of nearby $^{238}U$ nuclei, which means that some small part of the $^{238}U$ is "burned-up" in all nuclear fuels, especially in fast breeder reactors that operate with higher-energy neutrons. That same fast-fission effect is used to augment the energy released by modern thermonuclear weapons, by jacketing the weapon with $^{238}U$ to react with neutrons released by nuclear fusion at the center of the device. But the explosive effects of nuclear fission chain reactions can be reduced by using substances like moderators which slow down the speed of secondary neutrons.

## Fission Reactors

The cooling towers of the Philippsburg Nuclear Power Plant, in Germany.

Critical fission reactors are the most common type of nuclear reactor. In a critical fission reactor, neutrons produced by fission of fuel atoms are used to induce yet more fissions, to sustain a controllable amount of energy release. Devices that produce engineered but non-self-sustaining fission reactions are subcritical fission reactors. Such devices use radioactive decay or particle accelerators to trigger fissions.

Critical fission reactors are built for three primary purposes, which typically involve different engineering trade-offs to take advantage of either the heat or the neutrons produced by the fission chain reaction:

- *power reactors* are intended to produce heat for nuclear power, either as part of a generating station or a local power system such as a nuclear submarine.
- *research reactors* are intended to produce neutrons and/or activate radioactive sources for scientific, medical, engineering, or other research purposes.
- *breeder reactors* are intended to produce nuclear fuels in bulk from more abundant isotopes. The better known fast breeder reactor makes $^{239}Pu$ (a nuclear fuel) from the naturally very abundant $^{238}U$ (not a nuclear fuel). Thermal breeder reactors previously tested using $^{232}Th$ to breed the fissile isotope $^{233}U$ (thorium fuel cycle) continue to be studied and developed.

While, in principle, all fission reactors can act in all three capacities, in practice the tasks lead to conflicting engineering goals and most reactors have been built with only one of the above tasks in mind. (There are several early counter-examples, such as the HanfordN reactor, now decommissioned). Power reactors generally convert the kinetic energy of fission products into heat, which is used to heat a working fluid and drive a heat engine that generates mechanical or electrical power. The working fluid is usually water with a steam turbine, but some designs use other materials such as gaseous helium. Research reactors produce neutrons that are used in various ways, with the heat of fission being treated as an unavoidable waste product. Breeder reactors are a specialized form of research reactor, with the caveat that the sample being irradiated is usually the fuel itself, a mixture of $^{238}U$ and $^{235}U$. For a more detailed description of the physics and operating principles of critical fission reactors. For a description of their social, political, and environmental aspects.

## Fission Bombs

The mushroom cloud of the atomic bomb dropped on Nagasaki, Japan on August 9, 1945, rose over 18 kilometres (11 mi) above the bomb's hypocenter. An estimated 39,000 people were killed by the atomic bomb, of whom 23,145–28,113 were Japanese factory workers, 2,000 were Korean slave laborers, and 150 were Japanese combatants.

One class of nuclear weapon, a *fission bomb*, otherwise known as an *atomic bomb* or *atom bomb*, is a fission reactor designed to liberate as much energy as possible as rapidly as possible, before the released energy causes the reactor to explode (and the chain reaction to stop). Development of nuclear weapons was the motivation behind early research into nuclear fission which the Manhattan Project during World War II (September 1, 1939 – September 2, 1945) carried out most of the early scientific work on fission chain reactions, culminating in the three events involving fission bombs that occurred during the war. The first fission bomb, codenamed "The Gadget", was detonated during the Trinity Test in the desert of New Mexico on July 16, 1945. Two other fission bombs, codenamed "Little Boy" and "Fat Man", were used in combat against the Japanese cities of Hiroshima and Nagasaki in on August 6 and 9, 1945 respectively.

Even the first fission bombs were thousands of times more explosive than a comparable mass of chemical explosive. For example, Little Boy weighed a total of about four tons (of which 60 kg was nuclear fuel) and was 11 feet (3.4 m) long; it also yielded an explosion equivalent to about 15 kilotons of TNT, destroying a large part of the city of Hiroshima. Modern nuclear weapons (which include a thermonuclear *fusion* as well as one or more fission stages) are hundreds of times more energetic for their weight than the first pure fission atomic bombs, so that a modern single missile warhead bomb weighing less than 1/8 as much as Little Boy has a yield of 475,000 tons of TNT, and could bring destruction to about 10 times the city area.

While the fundamental physics of the fission chain reaction in a nuclear weapon is similar to the physics of a controlled nuclear reactor, the two types of device must be engineered quite differently. A nuclear bomb is designed to release all its energy at once, while a reactor is designed to generate a steady supply of useful power. While overheating of a reactor can lead to, and has led to, meltdown and steam explosions, the much lower uranium enrichment makes it impossible for a nuclear reactor to explode with the same destructive power as a nuclear weapon. It is also difficult to extract useful power from a nuclear bomb, although at least one rocket propulsion system, Project Orion, was intended to work by exploding fission bombs behind a massively padded and shielded spacecraft.

The strategic importance of nuclear weapons is a major reason why the technology of nuclear fission is politically sensitive. Viable fission bomb designs are, arguably, within the capabilities of many, being relatively simple from an engineering viewpoint. However, the difficulty of obtaining fissile nuclear material to realize the designs is the key to the relative unavailability of nuclear weapons to all but modern industrialized governments with special programs to produce fissile materials.

## History

### Discovery of Nuclear Fission

The discovery of nuclear fission occurred in 1938 in the buildings of Kaiser Wilhelm Society for Chemistry, today part of the Free University of Berlin, following nearly five decades of work on the science of radioactivity and the elaboration of new nuclear physics that described the components of atoms. In 1911, Ernest Rutherford proposed a model of the atom in which a very small, dense and positively charged nucleus of protons (the neutron had not yet been discovered) was surrounded by orbiting, negatively charged electrons (the Rutherford model).Niels Bohr improved upon this in 1913 by reconciling the quantum behavior of electrons (the Bohr model). Work by Henri Becquerel, Marie Curie, Pierre Curie, and Rutherford further elaborated that the nucleus, though tightly bound, could undergo different forms of radioactive decay, and thereby transmute into other elements. (For example, by alpha decay: the emission of an alpha particle—two protons and two neutrons bound together into a particle identical to a helium nucleus).

Some work in nuclear transmutation had been done. In 1917, Rutherford was able to accomplish transmutation of nitrogen into oxygen, using alpha particles directed at nitrogen $^{14}N + \alpha \rightarrow {}^{17}O + p$. This was the first observation of a nuclear reaction, that is, a reaction in which particles from one decay are used to transform another atomic nucleus. Eventually, in 1932, a fully artificial nuclear reaction and nuclear transmutation was achieved by Rutherford's colleagues Ernest Walton and John Cockcroft, who used artificially accelerated protons against lithium-7, to split this nucleus into two alpha particles. The feat was popularly known as "splitting the atom", although it was not the modern nuclear fission reaction later discovered in heavy elements, which is discussed below. Meanwhile, the possibility of *combining* nuclei—nuclear fusion—had been studied in connection with understanding the processes which power stars. The first artificial fusion reaction had been achieved by Mark Oliphant in 1932, using two accelerated deuterium nuclei (each consisting of a single proton bound to a single neutron) to create a helium-3 nucleus.

After English physicist James Chadwick discovered the neutron in 1932,Enrico Fermi and his colleagues in Rome studied the results of bombarding uranium with neutrons in 1934. Fermi concluded that his experiments had created new elements with 93 and 94 protons, which the group dubbed ausonium and hesperium. However, not all were convinced by Fermi's analysis of his results. The German chemist Ida Noddack notably suggested in print in 1934 that instead of creating a new, heavier element 93, that "it is conceivable that the nucleus breaks up into several large fragments." However, Noddack's conclusion was not pursued at the time.

The experimental apparatus with which Otto Hahn and Fritz Strassmann discovered nuclear fission in 1938

After the Fermi publication, Otto Hahn, Lise Meitner, and Fritz Strassmann began performing similar experiments in Berlin. Meitner, an Austrian Jew, lost her citizenship with the "Anschluss", the occupation and annexation of Austria into Nazi Germany in March 1938, but she fled in July 1938 to Sweden and started a correspondence by mail with Hahn in Berlin. By coincidence, her nephew Otto Robert Frisch, also a refugee, was also in Sweden when Meitner received a letter from

Hahn dated 19 December describing his chemical proof that some of the product of the bombardment of uranium with neutrons was barium. Hahn suggested a *bursting* of the nucleus, but he was unsure of what the physical basis for the results were. Barium had an atomic mass 40% less than uranium, and no previously known methods of radioactive decay could account for such a large difference in the mass of the nucleus. Frisch was skeptical, but Meitner trusted Hahn's ability as a chemist. Marie Curie had been separating barium from radium for many years, and the techniques were well-known. According to Frisch:

Was it a mistake? No, said Lise Meitner; Hahn was too good a chemist for that. But how could barium be formed from uranium? No larger fragments than protons or helium nuclei (alpha particles) had ever been chipped away from nuclei, and to chip off a large number not nearly enough energy was available. Nor was it possible that the uranium nucleus could have been cleaved right across. A nucleus was not like a brittle solid that can be cleaved or broken; George Gamow had suggested early on, and Bohr had given good arguments that a nucleus was much more like a liquid drop. Perhaps a drop could divide itself into two smaller drops in a more gradual manner, by first becoming elongated, then constricted, and finally being torn rather than broken in two? We knew that there were strong forces that would resist such a process, just as the surface tension of an ordinary liquid drop tends to resist its division into two smaller ones. But nuclei differed from ordinary drops in one important way: they were electrically charged, and that was known to counteract the surface tension.

The charge of a uranium nucleus, we found, was indeed large enough to overcome the effect of the surface tension almost completely; so the uranium nucleus might indeed resemble a very wobbly unstable drop, ready to divide itself at the slightest provocation, such as the impact of a single neutron. But there was another problem. After separation, the two drops would be driven apart by their mutual electric repulsion and would acquire high speed and hence a very large energy, about 200 MeV in all; where could that energy come from? ...Lise Meitner... worked out that the two nuclei formed by the division of a uranium nucleus together would be lighter than the original uranium nucleus by about one-fifth the mass of a proton. Now whenever mass disappears energy is created, according to Einstein's formula $E= mc^2$, and one-fifth of a proton mass was just equivalent to 200 MeV. So here was the source for that energy; it all fitted.

In short, Meitner and Frisch had correctly interpreted Hahn's results to mean that the nucleus of uranium had split roughly in half. Frisch suggested the process be named "nuclear fission," by analogy to the process of living cell division into two cells, which was then called binary fission. Just as the term nuclear "chain reaction" would later be borrowed from chemistry, so the term "fission" was borrowed from biology.

On 22 December 1938, Hahn and Strassmann sent a manuscript to *Naturwissenschaften* reporting that they had discovered the element barium after bombarding uranium with neutrons. Simultaneously, they communicated these results to Meitner in Sweden. She and Frisch correctly interpreted the results as evidence of nuclear fission. Frisch confirmed this experimentally on 13 January 1939. For proving that the barium resulting from his bombardment of uranium with neutrons was the product of nuclear fission, Hahn was awarded the Nobel Prize for Chemistry in 1944 (the sole recipient) "for his discovery of the fission of heavy nuclei". (The award was actually given to Hahn in 1945, as "the Nobel Committee for Chemistry decided that none of the year's nominations met the criteria as outlined in the will of Alfred Nobel." In such cases, the Nobel Foundation's statutes permit that year's prize be reserved until the following year).

German stamp honoring Otto Hahn and his discovery of nuclear fission (1979)

News spread quickly of the new discovery, which was correctly seen as an entirely novel physical effect with great scientific—and potentially practical—possibilities. Meitner's and Frisch's interpretation of the discovery of Hahn and Strassmann crossed the Atlantic Ocean with Niels Bohr, who was to lecture at Princeton University. I.I. Rabi and Willis Lamb, two Columbia University physicists working at Princeton, heard the news and carried it back to Columbia. Rabi said he told Enrico Fermi; Fermi gave credit to Lamb. Bohr soon thereafter went from Princeton to Columbia to see Fermi. Not finding Fermi in his office, Bohr went down to the cyclotron area and found Herbert L. Anderson. Bohr grabbed him by the shoulder and said: "Young man, let me explain to you about something new and exciting in physics." It was clear to a number of scientists at Columbia that they should try to detect the energy released in the nuclear fission of uranium from neutron bombardment. On 25 January 1939, a Columbia University team conducted the first nuclear fission experiment in the United States, which was done in the basement of Pupin Hall; the members of the team were Herbert L. Anderson, Eugene T. Booth, John R. Dunning, Enrico Fermi, G. Norris Glasoe, and Francis G. Slack. The experiment involved placing uranium oxide inside of an ionization chamber and irradiating it with neutrons, and measuring the energy thus released. The results confirmed that fission was occurring and hinted strongly that it was the isotope uranium 235 in particular that was fissioning. The next day, the Fifth Washington Conference on Theoretical Physics began in Washington, D.C. under the joint auspices of the George Washington University and the Carnegie Institution of Washington. There, the news on nuclear fission was spread even further, which fostered many more experimental demonstrations.

During this period the Hungarian physicist LeóSzilárd, who was residing in the United States at the time, realized that the neutron-driven fission of heavy atoms could be used to create a nuclear chain reaction. Such a reaction using neutrons was an idea he had first formulated in 1933, upon reading Rutherford's disparaging remarks about generating power from his team's 1932 experiment using protons to split lithium. However, Szilárd had not been able to achieve a neutron-driven chain reaction with neutron-rich light atoms. In theory, if in a neutron-driven chain reaction the number of secondary neutrons produced was greater than one, then each such reaction could trigger multiple additional reactions, producing an exponentially increasing number of reactions. It was thus a possibility that the fission of uranium could yield vast amounts of energy for civilian or military purposes (i.e., electric power generation or atomic bombs).

Szilard now urged Fermi (in New York) and Frédéric Joliot-Curie (in Paris) to refrain from publishing on the possibility of a chain reaction, lest the Nazi government become aware of the possibilities on the eve of what would later be known as World War II. With some hesitation Fermi agreed to self-censor. But Joliot-Curie did not, and in April 1939 his team in Paris, including Hans von Halban and Lew Kowarski, reported in the journal *Nature* that the number of neutrons emitted with nuclear fission of $^{235}U$ was then reported at 3.5 per fission. (They later corrected this to 2.6 per fission.) Simultaneous work by Szilard and Walter Zinn confirmed these results. The results suggested the possibility of building nuclear reactors (first called "neutronic reactors" by Szilard and Fermi) and even nuclear bombs. However, much was still unknown about fission and chain reaction systems.

## Fission Chain Reaction Realized

Drawing of the first artificial reactor, Chicago Pile-1.

"Chain reactions" at that time were a known phenomenon in *chemistry*, but the analogous process in nuclear physics, using neutrons, had been foreseen as early as 1933 by Szilárd, although Szilárd at that time had no idea with what materials the process might be initiated. Szilárd considered that neutrons would be ideal for such a situation, since they lacked an electrostatic charge.

With the news of fission neutrons from uranium fission, Szilárd immediately understood the possibility of a nuclear chain reaction using uranium. In the summer, Fermi and Szilard proposed the idea of a nuclear reactor (pile) to mediate this process. The pile would use natural uranium as fuel. Fermi had shown much earlier that neutrons were far more effectively captured by atoms if they were of low energy (so-called "slow" or "thermal" neutrons), because for quantum reasons it made the atoms look like much larger targets to the neutrons. Thus to slow down the secondary neutrons released by the fissioning uranium nuclei, Fermi and Szilard proposed a graphite "moderator," against which the fast, high-energy secondary neutrons would collide, effectively slowing them down. With enough uranium, and with pure-enough graphite, their "pile" could theoretically sustain a slow-neutron chain reaction. This would result in the production of heat, as well as the creation of radioactive fission products.

In August 1939, Szilard and fellow Hungarian refugees physicists Teller and Wigner thought that the Germans might make use of the fission chain reaction and were spurred to attempt to attract the attention of the United States government to the issue. Towards this, they persuaded German-Jewish refugee Albert Einstein to lend his name to a letter directed to President Franklin Roosevelt. The Einstein–Szilárd letter suggested the possibility of a uranium bomb deliverable by ship, which would destroy "an entire harbor and much of the surrounding countryside." The President received the letter on 11 October 1939 — shortly after World War II began in Europe, but two years before U.S. entry into it. Roosevelt ordered that a scientific committee be authorized for overseeing uranium work and allocated a small sum of money for pile research.

In England, James Chadwick proposed an atomic bomb utilizing natural uranium, based on a paper by Rudolf Peierls with the mass needed for critical state being 30–40 tons. In America, J. Robert Oppenheimer thought that a cube of uranium deuteride 10 cm on a side (about 11 kg of uranium) might "blow itself to hell." In this design it was still thought that a moderator would need to be used for nuclear bomb fission (this turned out not to be the case if the fissile isotope was separated). In December, Werner Heisenberg delivered a report to the German Ministry of War on the possibility of a uranium bomb. Most of these models were still under the assumption that the bombs would be powered by slow neutron reactions—and thus be similar to a reactor undergoing a meltdown.

In Birmingham, England, Frisch teamed up with Peierls, a fellow German-Jewish refugee. They had the idea of using a purified mass of the uranium isotope $^{235}U$, which had a cross section just determined, and which was much larger than that of $^{238}U$ or natural uranium (which is 99.3% the latter isotope). Assuming that the cross section for fast-neutron fission of $^{235}U$ was the same as for slow neutron fission, they determined that a pure $^{235}U$ bomb could have a critical mass of only 6 kg instead of tons, and that the resulting explosion would be tremendous. (The amount actually turned out to be 15 kg, although several times this amount was used in the actual uranium (Little Boy) bomb). In February 1940 they delivered the Frisch–Peierls memorandum. Ironically, they were still officially considered "enemy aliens" at the time. Glenn Seaborg, Joseph W. Kennedy, Arthur Wahl, and Italian-Jewish refugee Emilio Segrè shortly thereafter discovered $^{239}Pu$ in the decay products of $^{239}U$ produced by bombarding $^{238}U$ with neutrons, and determined it to be a fissile material, like $^{235}U$.

The possibility of isolating uranium-235 was technically daunting, because uranium-235 and uranium-238 are chemically identical, and vary in their mass by only the weight of three neutrons. However, if a sufficient quantity of uranium-235 could be isolated, it would allow for a fast neutron fission chain reaction. This would be extremely explosive, a true "atomic bomb." The discovery that plutonium-239 could be produced in a nuclear reactor pointed towards another approach to a fast neutron fission bomb. Both approaches were extremely novel and not yet well understood, and there was considerable scientific skepticism at the idea that they could be developed in a short amount of time.

On June 28, 1941, the Office of Scientific Research and Development was formed in the U.S. to mobilize scientific resources and apply the results of research to national defense. In September, Fermi assembled his first nuclear "pile" or reactor, in an attempt to create a slow neutron-induced chain reaction in uranium, but the experiment failed to achieve criticality, due to lack of proper materials, or not enough of the proper materials which were available.

Producing a fission chain reaction in natural uranium fuel was found to be far from trivial. Early nuclear reactors did not use isotopically enriched uranium, and in consequence they were required to use large quantities of highly purified graphite as neutron moderation materials. Use of ordinary water (as opposed to heavy water) in nuclear reactors requires enriched fuel — the partial separation and relative enrichment of the rare $^{235}U$ isotope from the far more common $^{238}U$ isotope. Typically, reactors also require inclusion of extremely chemically pure neutron moderator materials such as deuterium (in heavy water), helium, beryllium, or carbon, the latter usually as graphite. (The high purity for carbon is required because many chemical impurities such as the boron-10 component of natural boron, are very strong neutron absorbers and thus poison the chain reaction and end it prematurely).

Production of such materials at industrial scale had to be solved for nuclear power generation and weapons production to be accomplished. Up to 1940, the total amount of uranium metal produced in the USA was not more than a few grams, and even this was of doubtful purity; of metallic beryllium not more than a few kilograms; and concentrated deuterium oxide (heavy water) not more than a few kilograms. Finally, carbon had never been produced in quantity with anything like the purity required of a moderator.

The problem of producing large amounts of high purity uranium was solved by Frank Spedding using the thermite or "Ames" process. Ames Laboratory was established in 1942 to produce the large amounts of natural (unenriched) uranium metal that would be necessary for the research to come. The critical nuclear chain-reaction success of the Chicago Pile-1 (December 2, 1942) which used unenriched (natural) uranium, like all of the atomic "piles" which produced the plutonium for the atomic bomb, was also due specifically to Szilard's realization that very pure graphite could be used for the moderator of even natural uranium "piles". In wartime Germany, failure to appreciate the qualities of very pure graphite led to reactor designs dependent on heavy water, which in turn was denied the Germans by Allied attacks in Norway, where heavy water was produced. These difficulties—among many others— prevented the Nazis from building a nuclear reactor capable of criticality during the war, although they never put as much effort as the United States into nuclear research, focusing on other technologies.

## Manhattan Project and Beyond

In the United States, an all-out effort for making atomic weapons was begun in late 1942. This work was taken over by the U.S. Army Corps of Engineers in 1943, and known as the Manhattan Engineer District. The top-secret Manhattan Project, as it was colloquially known, was led by General Leslie R. Groves. Among the project's dozens of sites were: Hanford Site in Washington state, which had the first industrial-scale nuclear reactors; Oak Ridge, Tennessee, which was primarily concerned with uranium enrichment; and Los Alamos, in New Mexico, which was the scientific hub for research on bomb development and design. Other sites, notably the Berkeley Radiation Laboratory and the Metallurgical Laboratory at the University of Chicago, played important contributing roles. Overall scientific direction of the project was managed by the physicist J. Robert Oppenheimer.

In July 1945, the first atomic explosive device, dubbed "Trinity", was detonated in the New Mexico desert. It was fueled by plutonium created at Hanford. In August 1945, two more atomic devices – "Little Boy", a uranium-235 bomb, and "Fat Man", a plutonium bomb – were used against the Japanese cities of Hiroshima and Nagasaki.

In the years after World War II, many countries were involved in the further development of nuclear fission for the purposes of nuclear reactors and nuclear weapons. The UK opened the first commercial nuclear power plant in 1956. By 2013, there were 437 reactors in 31 countries.

## Natural Fission Chain-reactors on Earth

Criticality in nature is uncommon. At three ore deposits at Oklo in Gabon, sixteen sites (the so-called Oklo Fossil Reactors) have been discovered at which self-sustaining nuclear fission took place approximately 2 billion years ago. Unknown until 1972 (but postulated by Paul Kuroda in 1956), when French physicist Francis Perrin discovered the Oklo Fossil Reactors, it was realized that nature had beaten humans to the punch. Large-scale natural uranium fission chain reactions, moderated by normal water, had occurred far in the past and would not be possible now. This ancient process was able to use normal water as a moderator only because 2 billion years before the present, natural uranium was richer in the shorter-lived fissile isotope $^{235}U$ (about 3%), than natural uranium available today (which is only 0.7%, and must be enriched to 3% to be usable in light-water reactors).

## Fission Reaction Mechanism

Let us first to understand the mechanism of nuclear fission process based on simple liquid drop picture of a nucleus.

Bohr and Wheeler explained the fission process considering the liquid-drop model, where nucleus resembles a drop of liquid in the respect that each constituent particle interacts equally with its nearest neighbors. In a liquid droplet, the total energy of the molecules inside the droplet is not sufficient enough to overcome the forces that hold the molecules in the form of a droplet. However, if an external force is applied so that it is set into oscillation, the system passes through a series of stages, of which the most significant ones are shown in the figure. This drop is at first spherical (S-1) and then elongated into an ellipsoid (S-2). Although the volume remains constant, the surface area has increased. But provided the volume energy exceeds the surface energy, the drop will return to its original form. However, if the deforming force is sufficiently large, the drop will acquire a shape similar to a dumb-bell as in S-3. In this state, the surface energy will generally exceed the volume energy, which provides the cohesive force for the liquid drop. Consequently, the drop will not return to its initial shape but will rather split into two droplets. These will, at first, be somewhat deformed, as in S-4, but finally they will become spherical (S-5).

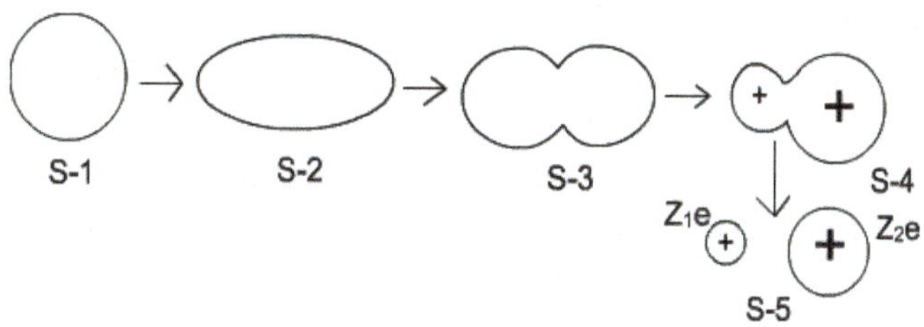


Schematic representation of nuclear fission process based on liquid drop model.

For further understanding, readers are suggested to go through literature.

Now let us calculate energy production in nuclear fission reaction and also establish basic equations related to power generation through this reaction.

Let us consider the following fission reaction

$$ {}_0^1n + {}_{92}^{235}U \rightarrow \left\lfloor {}_{92}^{236}U \right\rfloor \rightarrow {}_{54}^{140}Xe + {}_{38}^{94}Sr + {}_0^1n + Q $$

The fission fragments decay to their radioactive isotopes through β-decay in the following manner:

$$^{140}_{54}Xe \rightarrow ^{140}_{58}Ce + 4\beta^-$$

and

$$^{94}_{38}Sr \rightarrow ^{94}_{40}Zr + 2\beta^-$$

The total mass before fission is

$$^{235}_{92}U = 235.0439 amu$$
$$^{1}_{0}n = 1.0087 amu$$
$$Total = 236.0526 amu$$

The isotopic mass after fission is

$$^{140}_{58}Ce = 139.9054 amu$$
$$^{94}_{40}Zr = 93.9036 amu$$
$$2^{1}_{0}n = 2.0173 amu$$
$$6\beta^- = 0.0033 amu$$
$$Total = 235.8296 amu$$

Mass difference = 0.223 amu. Therefore, Q = 0.223 X 931 = 208 MeV.

In MKS unit, $Q = 208 \times 1.6 \times 10^{-13} = 3.32 \times 10^{-11}\ joules$.

Considering U of mass equal to 1 gm, the energy released is

$(6.02 \times 1023/235) \times 3.32 \times 10^{-11} = 8.50\ X\ 10^{10}\ joules$, which is indeed very large. In the unit of power, 1 gm U generates $8.50 \times 10^{10}$ watt -second i.e. $2.36 \times 10^{4}$ kw-hr. Therefore, one gm-atom U produces a power of $5.54 \times 10^{6}$ kw-hr.

Let us see how many fission reactions are required to generate 1 Watt power.

Example m3.3: When a U-235 nucleus undergoes fission, 200 MeV energy is released. How many fissions per sec are needed for generating power of 1 Watt?

Solution:

1 Watt = 1 joule per sec = $6.242 \times 10^{12}$ MeV/s

No. of fission needed per sec = $\frac{6.242 \times 10^{12}}{200} = 3.12 \times 10^{10}$

Next important issue related to the realization of generation of power from a reactor is to satisfy the condition of criticality. In a finite size reactor, some neutrons are lost by leaking out. The criticality condition is then defined as the effective multiplication factor, which is basically the ratio of number of neutrons resulting from fission in each generation to the total number lost by both absorption and leakage in the preceding generation. Let, for fissile material, which is a combination of U-238 and U-235, at an instant there be No fast neutrons available to produce fission. A few of them may produce fission in U-238 and hence the number of neutrons increases by a factor v and become Nov. These neutrons must be slowed down before they produce fission in U-235. The slowing down process is done in moderator (discussed later), which reduces the energy of the fast neutrons to thermal energy. Considering p to be the fraction slowed down, the available neutrons are Novp. Of these again, a fraction f may succeed in producing fission in U-235 before being lost by diffusion or otherwise. Therefore the number of U-235 nuclei undergoing fission is Novpf. If at each fission of U-235, fast neutrons are generated to start the cycle again, the total number of neutrons after one cycle is

$$N = N_0 vpfe$$

The ration N/No (=vpfe) is also known as reproduction factor, k. The requirement of criticality in a finite system is thus k = 1; in these circumstances a steady-state fission chain would be possible. If the conditions are such that k < 1, the chain would be convergent and gradually die out. In each generation, more neutrons would be lost in one way or another than are produced by fission, and so the neutron density, and hence the fission rate would decrease steadily. Such system is known to be subcritical. Finally, if k > 1, the chain is divergent and the system is known to be supercritical. More neutrons are produced than are lost in each generation, so that both neutron population and fission rate increase continuously.

How the power of a reactor is calculated?

Fission rate depends on two quantities, namely macroscopic cross section Σf for the process and flux Φ of monoenergetic neutrons.

$$Fission\ rate = \sum\nolimits_f \phi\, fissions/(cm)^3 (\sec)$$

Where

$$\sum\nolimits_f = N\sigma_f \ and\ \phi = nv$$

N is the number of fissile nuclei per cm3, and σf cm2 is the fission cross section; nis the neutron density, i.e., neutron per cm3, and v is the neutron velocity in cm2 per sec. In a reactor of volume V cm3, the fission rate will be VΣf Φ. Since it requires a fission rate of $3.1 \times 10^{10}$ fissions per sec to produce 1 watt of power, then the power of the nuclear reactor can be obtained using the following equation:

$$P = \frac{V\sum_f \phi}{3.1 \times 10^{10}} watts$$

If the mass of fissile material (U-235) is g, then we can write

$$g = \frac{235VN}{6.02 \times 10^{24}} grams -------- m$$

Therefore,

$$P = \frac{NV\sigma_f \phi}{3.1 \times 10^{10}} = 8.3 \times 10^{10} g\sigma_f \phi$$

If the mass of U-235 is expressed as w lb, then the above equation becomes,

$$P = 3.7 \times 10^{13} w \sigma_f \phi \ watts$$

Therefore, it is clear that the power of the reactor depends on the neutron flux. Therefore, the design of a nuclear reactor depends on efficient way of optimizing neutron number in the reactor.

## Basic Design Aspects of a Fission Reactor

The fundamental design aspect of a nuclear fission reactor involve following components in a reactor:

1. A reactor core,
2. Moderation stage,
3. Reflector covering the core region,
4. Coolant and energy removal,
5. Radiation shielding and
6. Overall control system.

Let us know the role of individual components of the reactor and their basic design aspects:

Reactor Core (S1):

In active core region, the fission chain is sustained and most of the energy of fission is released as heat. The core contains the nuclear fuel, consisting of a fissile nuclide and often a fertile material in addition. Among various reactor fuels, here we will discuss about uranium as a fuel material. The percentage of U is only 0.001 or less and hence extraction of metal would appear to be uneconomic.

The uranium to be used as a reactor fuel must be especially free from elements having appreciable cross-section for neutron capture. Most uranium ores contain about 1.5 to 10 lb of U3O8 per ton, i.e., roughly 0.1 to 0.5 % of U, and these are concentrated by a chemical process2. The ore is first leached with acid or alkali (sodium carbonate) depending on its nature. The uranium can be precipitated from the leach solution in one-way or another, but it has generally been found most satisfactory to remove the uranium by solvent extraction or ion-exchange techniques. Partially enriched uranium, containing from 1 to 5 % U-235 used as a fuel, is separated from its isotopes by the gaseous-diffusion method. In this method, isotopes of different molecular weights are separated through a porous barrier due to their different diffusion rates. The barriers used in the uranium separation process contain hundreds of millions of pores per square inch, the average pore diameter being about two-millionth of an inch. Based on the diffusion rate, which is inversely proportional to the molecular weight, a separation factor s.f. is defined in the following manner:

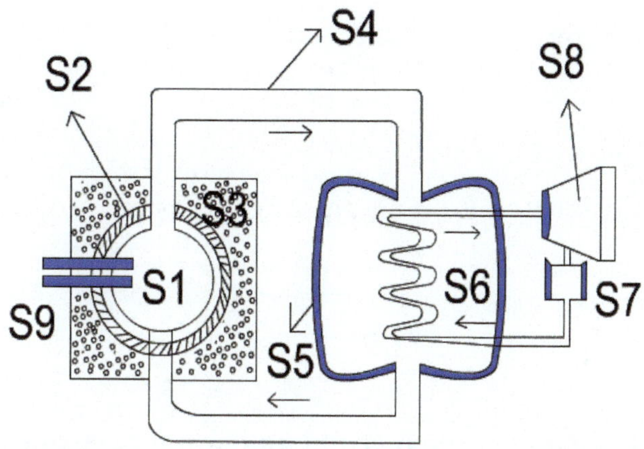


m3.2 Schematic of a nuclear fission reactor and various components.

$$s.f. = \sqrt{\frac{M(heavy)}{M(light)}}$$

Where M (heavy) and M (light) are the molecular weights of the heavier and lighter isotopic diffusion species, respectively. For the separation of the isotopes of uranium by diffusion, the gaseous compound employed is hexafluoride, UF6. This substance is a solid at ordinary temperatures, but it has relatively low sublimation temperature, so that it vaporizes readily. Taking uranium hexafluoride as the diffusion gas, M (heavy) is (238) + (6 x 19) = 352, whereas M (light) is (235) + (6 x 19) = 349. The theoretical separation factor is consequently,

$$s.f. = \sqrt{\frac{352}{349}} = 1.0043$$

The theoretical enrichment factor2 defined as s.f. - 1, is therefore 0.0043 for U-235 in UF6. A nuclear reactor cannot be used to release energy continuously unless the critical mass of the particular fuel, shape etc. is exceeded. For this reason, the power reactor must include extra fuel, and so it must obviously have excess reactivity to an appreciable extent at the time it commences operation.

Moderator (S2):

Most of the fissions result from the absorption of slow neutrons, and hence the presence of a moderator is must. The function of the moderator is to slow down the high-energy neutrons liberated in the fission reaction, mainly as a result of elastic scattering. The best moderators are materials consisting of elements of low mass number with low neutron capture probability. Examples of moderator are ordinary water, heavy water (deuterium oxide), beryllium oxide, carbon (as graphite), and hydrocarbons. For a given moderator material, there is a cross section, $\sigma s$, for neutron scattering by collision, and a cross-section, $\sigma a$ for neutron absorption. A parameter known as slowed down parameter, $\xi^2$ can be deduced from kinematics:

$$\xi = 1 + \frac{(A-1)^2}{2A} \ln \frac{A-1}{A+1}$$

where A is the atomic mass number.

This can also be defined as

$$\xi = \left( -\ln \frac{E_0}{E} \right)$$

where E = energy before collision and Eo = energy after collision. Moderating ratio (M.R.) can be defined as

$$\text{M.R.} = \frac{\sigma_s}{\sigma_a} \xi$$

The relative amounts and the nature of the fuel and moderator determine the energies of the most neutrons causing fission. Detailed graphitization process of petroleum coke and its use as moderator can be found in the literature.

Reflector (S3):

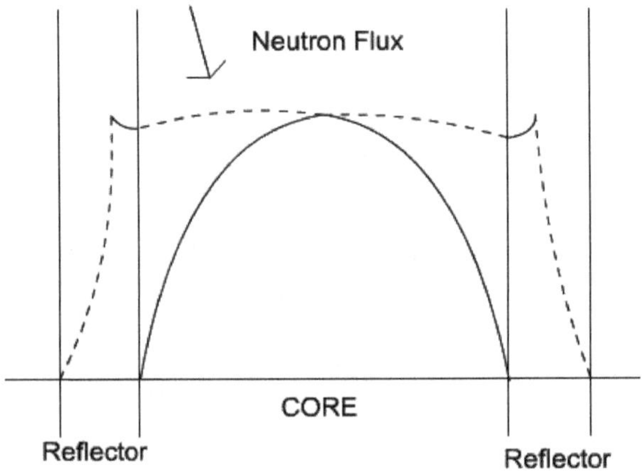

Thermal neutron flux distributions in bare and reflected reactors

Reflector is a material placed around the reactor core (that contains the fuel and the moderator), which prevents neutrons from escaping the reactor core. Therefore, reflector decreases the loss of neutrons from the core by scattering back many of those, which have escaped. Hence, the use of a reflector results in a decrease in the critical mass of the fissile nuclide. Good moderators are good reflectors in case where core contains moderator to slow down the neutrons. On the other hand, when it is required that most of the fissions should be caused by neutrons of high energy, the presence of moderating material must be avoided; the reflector then consists of a dense element of high mass number. For a thermal reactor, heavy water, beryllium (or its oxide), or graphite is generally used for this purpose. As many neutrons get reflected back from the reflector, the leakage of neutrons is reduced, which results in the decrease of the critical mass of fissile material as compared to bare reactor. In addition to decreasing the volume and mass of the critical core, the reflector also plays a role in the increase of the average power output of the reactor for a given mass of fuel and peak neutron flux. Because of the return of neutrons to the core by the reflector, the flux near the boundaries will be much greater than in the absence of a reflector, resulting in an increase of average neutron flux over the whole core, and hence the average power output for a given flux. The spatial distribution of neutrons for bare and reflected reactors is shown in the figure.

Coolant and Energy Removal (S4-S6):

The major portion of the heat generated in the core is due to the kinetic energy of the fission fragments, which usually manifests itself as heat release in the fuel elements. Heat is also produced from the slowing down of neutrons and beta particles, and the absorption of various gamma radiations. The heat so generated in a reactor is removed by circulation of suitable coolant and transported to heat exchanger. Examples of coolants are liquid water, liquid sodium, some organic compounds, and gases of air, carbon dioxide, helium etc. The choice of coolant, arrangement of fuel and coolant and method of heat removal are the primary considerations while designing a reactor. This is based on the engineering knowledge of heat transfer, hydrodynamics, and thermal stress etc. There are three general mechanisms whereby heat is transferred from one point to another, namely, conduction, convection and radiation. In conduction process, heat is transferred from one point to another by molecular interaction without any macroscopic displacement of matter. The flow of heat is governed by the Fourier equation

$$q = -kA\frac{dt}{dx}$$

Where q is the heat per unit time conducted along x direction through a plane of area A normal to this direction, and at a point where the temperature gradient is dt/dx. K is the thermal conductivity with unit Cal/(sec)(cm2)(oC/cm). Thermal conductivity is a physical property of the medium through which the heat conduction occurs. Considering k to be independent of temperature, the above equation can be integrated for unidirectional heat flow by conduction in a slab of constant cross section

$$q = -kA\frac{t_1 - t_2}{x_1 - x_2}$$

Where t1 and t2 are temperatures at two points whose coordinates are x1 and x2 respectively. Replacing t1-t2 by Δt, the temperature difference and x1-x2 by L, the length of the heat-flow path the above equation can be written as

$$q = -\frac{\Delta t}{\frac{L}{kA}}$$

This expression is analogous to Ohm's law, I = V/R; hence the quantity can be defined as thermal resistance for a conductor. The temperature difference Δt is analogous to potential difference V.

The convection process occurs due to the macroscopic motion of the medium, and is generally concerned with the transfer of heat across a solid-fluid interface. In free convection, the motion is influenced by the buoyant forces generated in the fluid due to temperature differences within it. In forced convection, on the other hand, the fluid is displaced by mechanical means, e.g., by a pump. The fundamental equation of convective heat transfer for both free and forced motion of the fluid is

$$q = hA_h\Delta t$$

Where q is the rate of heat transfer to or from a surface area Ah when the temperature difference is Δt. The quantity h is commonly known as heat-transfer coefficient or unit thermal conductance. The above equation applies to convective heat flow equation in either direction, i.e., from solid to fluid or from fluid to solid; the actual direction of the flow depends, of course, upon the sign of Δt. Rearranging the above equation, we get,

$$q = \frac{\Delta t}{\frac{1}{hA_k}}$$

defining 1/hAh as the thermal resistance of the convective heat transfer. As far as heat transfer is concerned, one can consider thermal circuits consisting of different stages of thermal conduction and convection. Let us first take an example of an infinite slab of finite thickness with no internal heat source.

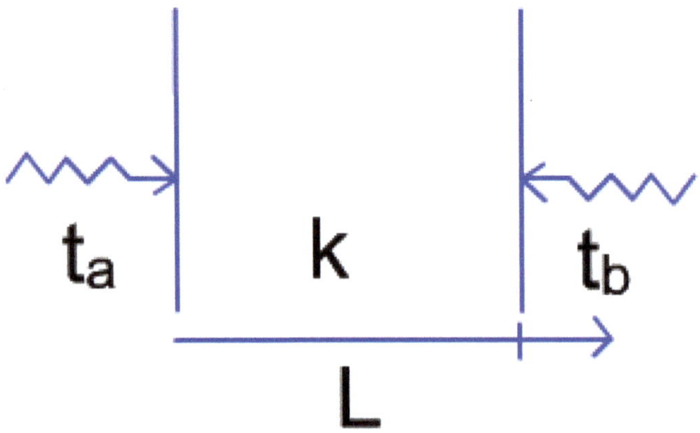


Heat conduction in infinite slab with convection boundaries

It is considered that heat is transferred at both faces by convection, and heat transfer coefficients are ha and hb respectively. Total thermal resistance of this thermal circuit can be written as

$$R = \frac{1}{h_a A} + \frac{L}{kA} + \frac{1}{h_b A} = \frac{1}{UA}$$

where A is the heat flow area and U is the overall coefficient of heat transfer having same dimension of heat transfer coefficient. Rate of heat flow for this system is

$$q = \frac{t_a - t_b}{R} = UA(t_a - t_b)$$

or

$$t_a - t_b = \frac{q}{UA}$$

The idealization of an infinite slab is a good approximation to real (finite) systems. In many cases coolant circuit has cylindrical or components of various other shapes. The simplest coolant circuit for heat-removal in a reactor is the direct-cycle one in which the coolant is passed once through the reactor and then to a sink. Sometimes river water is used as a coolant, and after passage through the reactor it returns to the river, which acts as a sink. In some cases, air is used as a coolant and the atmosphere as a sink. These systems, although simple in principle, are not without problems. River water must be purified before using it as coolant, and provision must be made for preventing the discharge of the radioactive air or water to the surroundings. For better control over the radioactive pollution, a secondary coolant system in conjunction with primary coolant circuit must be used. Heavy water, for example, is a desirable coolant from the standpoint of its nuclear properties, but due to its high cost, a secondary coolant, e.g., ordinary water, is required to transfer the heat to the sink. Finally, steam is produced from hot coolant and is converted to the mechanical energy of the turbine for production of electricity (S8).

## Shielding and Radiation Protection (S9)

One of the important aspects of the design of nuclear reactor and its associated equipments is the proper shielding for radiation protection, which will be discussed later. Shielding is done by attenuating nuclear radiation, mainly primary neutrons and gamma rays originating within the core itself and the secondary gamma rays produced by neutron interactions with material external to the core, e.g., reflector, coolant shield etc. The choice of shielding material and its proper thickness to bring down the radiation level to acceptable value is the important aspect of its design. Another aspect, which has to be taken care, is the temperature distribution in the shield material, as absorption of radiation energy will cause heating of the shield material. Therefore, knowledge of the distribution of neutrons and gamma rays throughout the thickness of the shield is very important. In order to get this knowledge, one approach is to determine the interaction cross sections and attenuation coefficients of neutrons and gamma rays in the shield material in analytical manner. However, due to the complex nature of the radiation source and its distribution in the reactor core and shield, and many interactions over a considerable range of energies, this approach is not practical. A semi-analytic procedure is possible, utilizing relatively simple but somewhat approximate mathematical expressions combined with experimental data of a statistical nature. The design of the reactor shield can be decided based on the following stages: first, the maximum permissible radiation dose rates at various locations must be established. This implies the knowledge of the physical layout of the reactor and its associated equipment. With the knowledge of the reactor system, a preliminary choice can be made of reactor materials and shield layout, so that the attenuation can be carried out. If for this the system data are available the comparison procedure can be used, otherwise a simple form of semi-analytic method is applied. The design process is thus iterative in nature, with the calculations becoming more and more refined as the final stages are approached.

Control System:

The basic purpose of a reactor control system is to provide smooth and steady operation. The control operation includes starting of the reactor, bringing power output to the desired level, maintain it at that level, and then shut down the reactor when necessary. A nuclear reactor cannot be used to release energy continuously unless the critical mass for the particular fuel, shape etc. is exceeded. For this reason, a power reactor must include extra fuel and so obviously have excess reactivity to an appreciable extent at the time it commences operation, which may lead to a rapid increase in neutron flux. Therefore, an essential aspect of reactor control lies in the precautions to be taken to prevent an excessive rate of increase in neutron flux when the power level is being raised. Since the power level of a reactor is proportional to the neutron flux, the obvious basis for reactor control is to vary the reproduction or multiplication factor or reactivity. If this factor is greater than unity, the reactor is supercritical and the power level will increase continuously. Upon decreasing the factor to unity, so that the reactor is just critical, the power output will remain constant (apart from transient changes) at the level attained at that time. Finally, by making the reactor subcritical, i.e., by reducing the effective multiplication factor to unity, the power level will be decreased. Multiplication factor can be varied by the addition or removal of fuel, moderator, reflector or a neutron absorber. Either individually or combining these methods has been used to control the operation of nuclear reactor. In thermal reactors, generally the insertion or withdrawal of materials (boron, steel, cadmium etc.) having large capture cross section of neutrons is being used in the control process. The control rods may be located within the core or in the reflector close to the core where the thermal neutron flux is high. Because of its high melting point and other useful properties, the element hafnium is employed in water-cooled reactors. This process is useful in controlling the reactors of low power output. However, this is not efficient for reactors having considerable reactivity with high neutron flux. An improvement in neutron economy can be achieved by combining motion of the core material, i.e., fuel and moderator, with that of an absorber. The lower portion of the coarse control rod may be constructed of the same material as the reactor core, whereas upper portion contains cadmium. When such a rod is lowered, so that neutron absorber is inserted, some of the core is removed at the same time, thereby bringing about a further decrease in reactivity. In a fast-neutron reactor, control by means of a neutron absorber is not generally satisfactory because of the low capture cross sections for neutrons of high energy. The reactivity can, however, be changed by the removal of fuel material from (or addition to) the core or by the movement of a part of the reflector. For the reactors of very low output powers, generally manual control is done, whereas otherwise the manual control is not feasible. Emergency safety action, for example, must be made automatic. If an instrument indicates that the neutron flux is rising at a dangerously rapid rate, or if the power level is appreciably higher than that for which the reactor was designed, an automatic system must be devised to tackle such situation. The control system consists of three loops, namely, operator loop, automatic loop and load loop. In 'operator loop', any information from any instrumental part of the reactor is received by the operator, who in turn, exerts appropriate action or conveys a signal to the control block. It is then the function of the control block to exert the desired control action on the reactor, usually with the expenditure of power from an external source. In the 'automatic loop', the information received from the reactor by the instruments is fed directly into the control block, the operator being by-passed. The 'load loop' is intended to represent the interaction of the nuclear aspects of the reactor, i.e., the fuel, moderator, and reflector, with the external conditions, or non-nuclear aspects. The external conditions may be regarded as constituting, in general sense, the 'load' on the reactor. The closed cycle in the block diagram implies that the load can affect the reactor, which in turn influences the load. The changes in the external conditions are indicated on suitable instruments, and information is fed back to the control block via operator and automatic loops so that the necessary action may be taken.

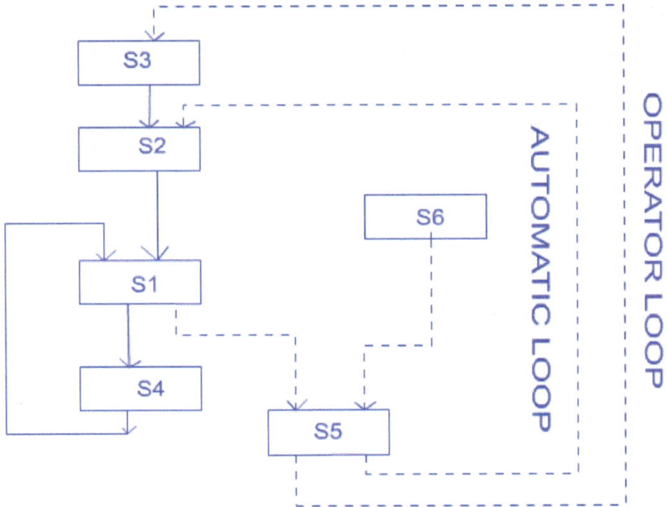


Block diagram of the reactor control system (Minor modification required)

## References


- H. Von Halban; F. Joliot & L. Kowarski (1939). "Number of Neutrons Liberated in the Nuclear Fission of Uranium". Nature. 143 (3625): 680. Bibcode:1939Natur.143..680V. doi:10.1038/143680a0
- "U.S. Nuclear Power Plants. General Statistical Information". Nuclear Energy Institute. Archived from the original on 22 October 2008. Retrieved 3 October 2009
- Tsetkov, Pavel; Usman, Shoaib (2011). Krivit, Steven, ed. Nuclear Energy Encyclopedia: Science, Technology, and Applications. Hoboken, NJ: Wiley. pp. 48; 85. ISBN 978-0-470-89439-2
- Shinn, E.; Et., al. (2013). "Nuclear energy conversion with stacks of graphene nanocapacitors". Complexity. doi:10.1002/cplx.21427
- "Cockcroft and Walton split lithium with high energy protons April 1932". Outreach.phy.cam.ac.uk. 1932-04-14. Retrieved 2013-01-04
- Kragh, Helge (1999). Quantum Generations: A History of Physics in the Twentieth Century. Princeton NJ: Princeton University Press. p. 286. ISBN 0-691-09552-3
- "DOE Fundamentals Handbook: Nuclear Physics and Reactor Theory" (PDF). US Department of Energy. Archived from the original (PDF) on 23 April 2008. Retrieved 24 September 2008
- E. Rutherford (1911). "The scattering of α and β particles by matter and the structure of the atom" (PDF). Philosophical Magazine. 21 (4): 669–688. Bibcode:2012PMag...92..379R. doi:10.1080/14786435.2011.617037
- Bernstein, Jeremy (2008). Nuclear Weapons: What You Need to Know. Cambridge University Press. p. 312. ISBN 978-0-521-88408-2. Retrieved 17 March 2011
- "International Scientific Journal for Alternative Energy and Ecology, DIRECT CONVERSION OF NUCLEAR ENERGY TO ELECTRICITY, Mark A. Prelas" (PDF)
- Golubev, V. I.; Dolgov, V. V.; Dulin, V. A.; Zvonarev, A. V.; Smetanin, É. Y.; Kochetkov, L. A.; Korobeinikov, V. V.; Liforov, V. G.; Manturov, G. N.; Matveenko, I. P.; Tsibulya, A. M. (1993). "Fast-reactor actinoid transmutation". Atomic Energy. 74: 83. doi:10.1007/BF00750983
- Nuke-Rebuke: Writers & Artists Against Nuclear Energy & Weapons (The Contemporary anthology series). The Spirit That Moves Us Press. May 1, 1984. pp. 22–29. ISBN 0930370155
- Juhasz, Albert J.; Rarick, Richard A.; Rangarajan, Rajmohan. "High Efficiency Nuclear Power Plants Using Liquid Fluoride Thorium Reactor Technology" (PDF). NASA. Retrieved 27 October 2014
- Tatsuichirō Akizuki; Gordon Honeycombe (March 1982). Nagasaki 1945: the first full-length eyewitness account of the atomic bomb attack on Nagasaki. Quartet Books. pp. 134–137. ISBN 978-0-7043-3382-6
- P. K. Kuroda (1956). "On the Nuclear Physical Stability of the Uranium Minerals" (PDF). The Journal of Chemical Physics. 25 (4): 781. Bibcode:1956JChPh..25..781K. doi:10.1063/1.1743058
- Lipper, Ilan; Stone, Jon. "Nuclear Energy and Society". University of Michigan. Archived from the original on 1 April 2009. Retrieved 3 October 2009
- Ida Noddack (1934). "Über das Element 93". Zeitschrift für Angewandte Chemie. 47 (37): 653. doi:10.1002/ange.19340473707
- Till, Charles. "Nuclear Reaction: Why Do Americans Fear Nuclear Power?". Public Broadcasting Service (PBS). Retrieved 9 November 2006
- The Impact of the A-bomb, Hiroshima and Nagasaki, 1945–85. Iwanami Shoten. 1 January 1985. pp. 56–78. ISBN 978-4-00-009766-6
- "Oklo: Natural Nuclear Reactors". Office of Civilian Radioactive Waste Management. Archived from the original on 16 March 2006. Retrieved 28 June 2006

6

# Radioactivity: Formulation and Theory

The process by which a nucleus can emit particles such as alpha, beta and gamma without being subjected to any external disturbance is called radioactivity. The process is governed by radioactive decay in which a nucleus disintegrates when it emits charged particles. In order to completely understand radioactivity, it is necessary to understand the processes related to it. The following section elucidates the varied processes and mechanisms associated with this area of study.

## Radioactive Decay

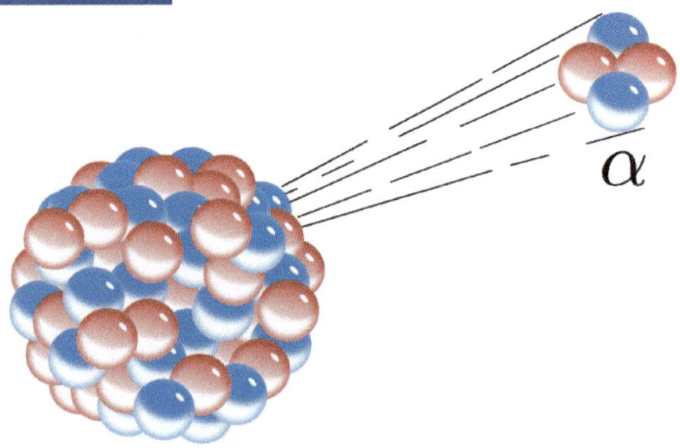


Alpha decay is one type of radioactive decay, in which an atomic nucleus emits an alpha particle, and thereby transforms (or "decays") into an atom with a mass number decreased by 4 and atomic number decreased by 2.

Radioactive decay (also known as nuclear decay or radioactivity) is the process by which an unstable atomic nucleus loses energy (in terms of mass in its rest frame) by emitting radiation, such as an alpha particle, beta particle with neutrino or only a neutrino in the case of electron capture, gamma ray, or electron in the case of Internal conversion. A material containing such unstable nuclei is considered radioactive. Certain highly excited short-lived nuclear states can decay through neutron emission, or more rarely, proton emission.

Radioactive decay is a stochastic (i.e. random) process at the level of single atoms, in that, according to quantum theory, it is impossible to predict when a particular atom will decay, regardless of how long the atom has existed. However, for a collection of atoms, the collection's expected decay rate is characterized in terms of their measured decay constants or half-lives. This is the basis of radiometric dating. The half-lives of radioactive atoms have no known upper limit, spanning a time range of over 55 orders of magnitude, from nearly instantaneous to far longer than the age of the universe.

A radioactive nucleus with zero spin can have no defined orientation, and hence emits the total momentum of its decay products isotropically (all directions and without bias). If there are multiple particles produced during a single decay, as in beta decay, their *relative* angular distribution, or spin directions may not be isotropic. Decay products from a nucleus with spin may be distributed non-isotropically with respect to that spin direction, either because of an external influence such as an electromagnetic field, or because the nucleus was produced in a dynamic process that constrained the direction of its spin. Such a parent process could be a previous decay, or a nuclear reaction.

The decaying nucleus is called the *parent radionuclide* (or *parent radioisotope*), and the process produces at least one *daughter nuclide*. Except for gamma decay or internal conversion from a nuclear excited state, the decay is a nuclear transmutation resulting in a daughter containing a different number of protons or neutrons (or both). When the number of protons changes, an atom of a different chemical element is created.

The first decay processes to be discovered were alpha decay, beta decay, and gamma decay. Alpha decay occurs when the nucleus ejects an alpha particle (helium nucleus). This is the most common process of emitting nucleons, but highly excited nuclei can eject single nucleons, or in the case of cluster decay, specific light nuclei of other elements. Beta decay occurs when the nucleus emits an electron or positron and a neutrino, in a process that changes a proton to a neutron or the converse. Highly excited neutron-rich nuclei, formed as the product of other types of decay, occasionally lose energy by way of neutron emission, resulting in a change from one isotope to another of the same element. The nucleus may capture an orbiting electron, causing a proton to convert into a neutron in a process called electron capture. All of these processes result in a well-defined nuclear transmutation.

By contrast, there are radioactive decay processes that do not result in a nuclear transmutation. The energy of an excited nucleus may be emitted as a gamma ray in a process called gamma decay, or that energy may be lost when the nucleus interacts with an orbital electron causing its ejection from the atom, in a process called internal conversion.

Another type of radioactive decay results in products that vary, appearing as two or more "fragments" of the original nucleus with a range of possible masses. This decay, called spontaneous fission, happens when a large unstable nucleus spontaneously splits into two (or occasionally three) smaller daughter nuclei, and generally leads to the emission of gamma rays, neutrons, or other particles from those products.

For a summary table showing the number of stable and radioactive nuclides in each category. There are 29 naturally occurring chemical elements on Earth that are radioactive. They are those that contain 34 radionuclides that date before the time of formation of the solar system, and are known as primordial nuclides. Well-known examples are uranium and thorium, but also included are naturally occurring long-lived radioisotopes, such as potassium-40. Another 50 or so shorter-lived radionuclides, such as radium and radon, found on Earth, are the products of decay chains that began with the primordial nuclides, or are the product of ongoing cosmogenic processes, such as the production of carbon-14 from nitrogen-14 in the atmosphere by cosmic rays. Radionuclides may also be produced artificially in particle accelerators or nuclear reactors, resulting in 650 of these with half-lives of over an hour, and several thousand more with even shorter half-lives.

## History of Discovery

Radioactivity was discovered in 1896 by the French scientist Henri Becquerel, while working with phosphorescent materials. These materials glow in the dark after exposure to light, and he suspected that the glow produced in cathode ray tubes by X-rays might be associated with phosphorescence. He wrapped a photographic plate in black paper and placed various phosphorescent salts on it. All results were negative until he used uranium salts. The uranium salts caused a blackening of the plate in spite of the plate being wrapped in black paper. These radiations were given the name "Becquerel Rays".

Pierre and Marie Curie in their Paris laboratory, before 1907

It soon became clear that the blackening of the plate had nothing to do with phosphorescence, as the blackening was also produced by non-phosphorescent salts of uranium and metallic uranium. It became clear from these experiments that there was a form of invisible radiation that could pass through paper and was causing the plate to react as if exposed to light.

At first, it seemed as though the new radiation was similar to the then recently discovered X-rays. Further research by Becquerel, Ernest Rutherford, Paul Villard, Pierre Curie, Marie Curie, and others showed that this form of radioactivity was significantly more complicated. Rutherford was the first to realize that all such elements decay in accordance with the same mathematical exponential formula. Rutherford and his student Frederick Soddy were the first to realize that many decay processes resulted in the transmutation of one element to another. Subsequently, the radioactive displacement law of Fajans and Soddy was formulated to describe the products of alpha and beta decay.

The early researchers also discovered that many other chemical elements, besides uranium, have radioactive isotopes. A systematic search for the total radioactivity in uranium ores also guided Pierre and Marie Curie to isolate two new elements: polonium and radium. Except for the radioactivity of radium, the chemical similarity of radium to barium made these two elements difficult to distinguish.

Marie and Pierre Curie's study of radioactivity is an important factor in science and medicine. After their research on Becquerel's rays led them to the discovery of both radium and polonium, they coined the term "radioactivity". Their research on the penetrating rays in uranium and the discovery of radium launched an era of using radium for the treatment of cancer. Their exploration of radium could be seen as the first peaceful use of nuclear energy and the start of modern nuclear medicine.

## Early Health Dangers

Taking an X-ray image with early Crookes tube apparatus in 1896. The Crookes tube is visible in the centre. The standing man is viewing his hand with a fluoroscope screen; this was a common way of setting up the tube. No precautions against radiation exposure are being taken; its hazards were not known at the time.

The dangers of ionizing radiation due to radioactivity and X-rays were not immediately recognized.

### X-rays

The discovery of xrays by Wilhelm Röntgen in 1895 led to widespread experimentation by scientists, physicians, and inventors. Many people began recounting stories of burns, hair loss and worse in technical journals as early as 1896. In February of that year, Professor Daniel and Dr. Dudley of Vanderbilt University performed an experiment involving X-raying Dudley's head that resulted in his hair loss. A report by Dr. H.D. Hawks, of his suffering severe hand and chest burns in an X-ray demonstration, was the first of many other reports in *Electrical Review*.

Other experimenters, including Elihu Thomson and Nikola Tesla, also reported burns. Thomson deliberately exposed a finger to an X-ray tube over a period of time and suffered pain, swelling, and blistering. Other effects, including ultraviolet rays and ozone, were sometimes blamed for the damage, and many physicians still claimed that there were no effects from X-ray exposure at all.

Despite this, there were some early systematic hazard investigations, and as early as 1902 William Herbert Rollins wrote almost despairingly that his warnings about the dangers involved in the careless use of X-rays was not being heeded, either by industry or by his colleagues. By this time, Rollins had proved that X-rays could kill experimental animals, could cause a pregnant guinea pig to abort, and that they could kill a fetus. He also stressed that "animals vary in susceptibility to the external action of X-light" and warned that these differences be considered when patients were treated by means of X-rays.

### Radioactive Substances

However, the biological effects of radiation due to radioactive substances were less easy to gauge. This gave the opportunity for many physicians and corporations to market radioactive substances as patent medicines. Examples were radium enema treatments, and radium-containing waters to be drunk as tonics. Marie Curie protested against this sort of treatment, warning that the effects of radiation on the human body were not well understood. Curie later died from aplastic anaemia, likely caused by exposure to ionizing radiation. By the 1930s, after a number of cases of bone necrosis and death of radium treatment enthusiasts, radium-containing medicinal products had been largely removed from the market (radioactive quackery).

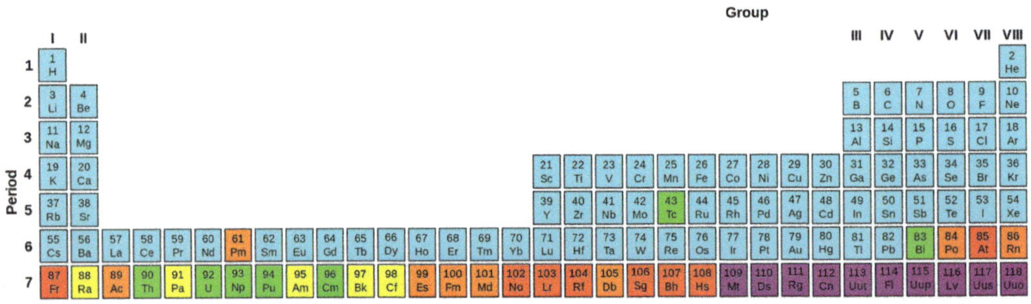


Radioactivity is characteristic of elements with large atomic number. Elements with at least one stable isotope are shown in light blue. Green shows elements whose most stable isotope has a half-life measured in millions of years. Yellow and orange are progressively less stable, with half-lives in thousands or hundreds of years, down toward one day. Red and purple show highly and extremely radioactive elements where the most stable isotopes exhibit half-lives measured on the order of one day and much less.

## Radiation Protection

Only a year after Röntgen's discovery of X rays, the American engineer Wolfram Fuchs (1896) gave what is probably the first protection advice, but it was not until 1925 that the first International Congress of Radiology (ICR) was held and considered establishing international protection standards. The effects of radiation on genes, including the effect of cancer risk, were recognized much later. In 1927, Hermann Joseph Muller published research showing genetic effects and, in 1946, was awarded the Nobel Prize in Physiology or Medicine for his findings.

The second ICR was held in Stockholm in 1928 and proposed the adoption of the rontgen unit, and the 'International X-ray and Radium Protection Committee' (IXRPC) was formed. Rolf Sievert was named Chairman, but a driving force was George Kaye of the British National Physical Laboratory. The committee met in 1931, 1934 and 1937.

After World War II, the increased range and quantity of radioactive substances being handled as a result of military and civil nuclear programmes led to large groups of occupational workers and the public being potentially exposed to harmful levels of ionising radiation. This was considered at the first post-war ICR convened in London in 1950, when the present International Commission on Radiological Protection (ICRP) was born. Since then the ICRP has developed the present international system of radiation protection, covering all aspects of radiation hazard.

## Units of Radioactivity

The International System of Units (SI) unit of radioactive activity is the becquerel (Bq), named in honour of the scientist Henri Becquerel. One Bq is defined as one transformation (or decay or disintegration) per second.

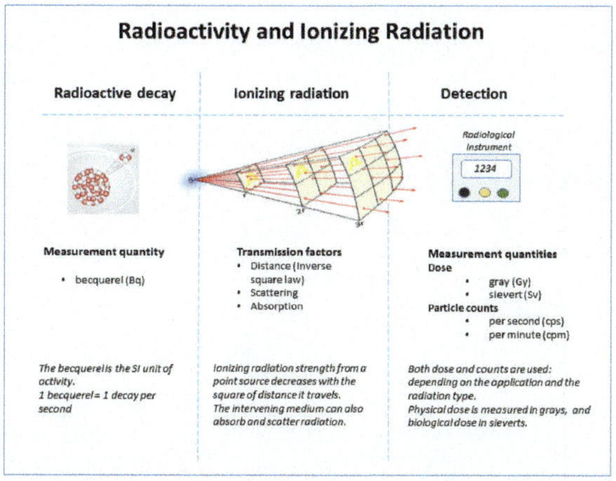


Graphic showing relationships between radioactivity and detected ionizing radiation

An older unit of radioactivity is the curie, Ci, which was originally defined as "the quantity or mass of radium emanation in equilibrium with one gram of radium (element)". Today, the curie is defined as $3.7\times10^{10}$ disintegrations per second, so that 1 curie (Ci) = $3.7\times10^{10}$ Bq. For radiological protection purposes, although the United States Nuclear Regulatory Commission permits the use of the unit curie alongside SI units, the European Union European units of measurement directives required that its use for "public health ... purposes" be phased out by 31 December 1985.

## Types of Decay

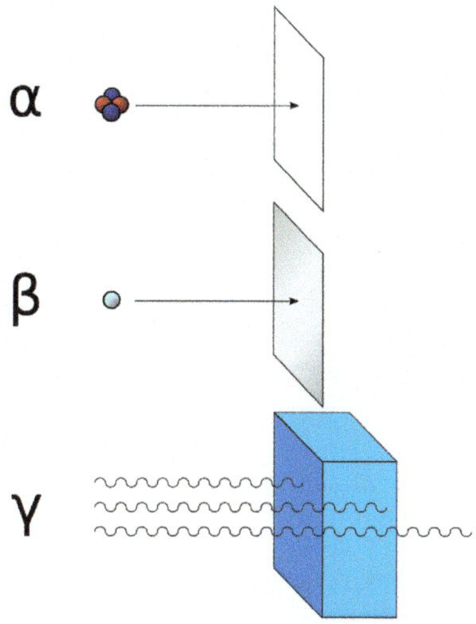


Alpha particles may be completely stopped by a sheet of paper, beta particles by aluminium shielding. Gamma rays can only be reduced by much more substantial mass, such as a very thick layer of lead.

Early researchers found that an electric or magnetic field could split radioactive emissions into three types of beams. The rays were given the names alpha, beta, and gamma, in order of their ability to penetrate matter. While alpha decay was observed only in heavier elements of atomic number 52 (tellurium) and greater, the other two types of decay were produced by all of the elements. Lead, atomic number 82, is the heaviest element to have any isotopes stable (to the limit of measurement) to radioactive decay. Radioactive decay is seen in all isotopes of all elements of atomic number 83 (bismuth) or greater. Bismuth, however, is only very slightly radioactive, with a half-life greater than the age of the universe; radioisotopes with extremely long half-lives are considered effectively stable for practical purposes.

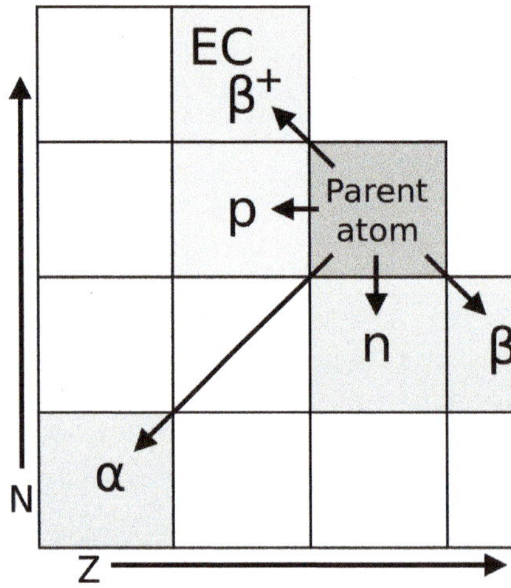


Transition diagram for decay modes of a radionuclide, with neutron number *N* and atomic number *Z* (shown are α, β±, p+, and n° emissions, EC denotes electron capture).

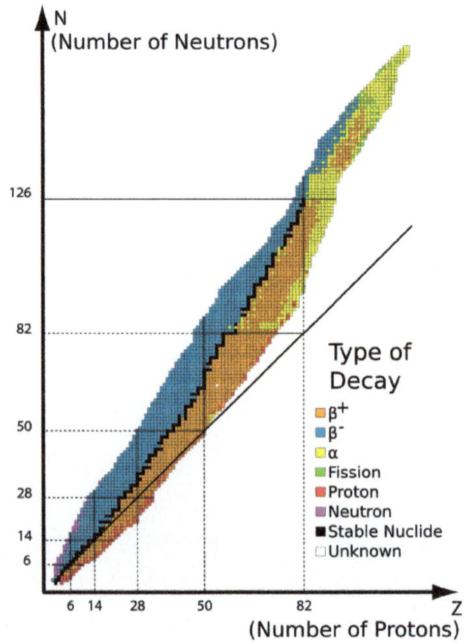


Types of radioactive decay related to N and Z numbers

In analysing the nature of the decay products, it was obvious from the direction of the electromagnetic forces applied to the radiations by external magnetic and electric fields that alpha particles carried a positive charge, beta particles carried a negative charge, and gamma rays were neutral. From the magnitude of deflection, it was clear that alpha particles were much more massive than beta particles. Passing alpha particles through a very thin glass window and trapping them in a discharge tube allowed researchers to study the emission spectrum of the captured particles, and ultimately proved that alpha particles are helium nuclei. Other experiments showed beta radiation, resulting from decay and cathode rays, were high-speed electrons. Likewise, gamma radiation and X-rays were found to be high-energy electromagnetic radiation.

The relationship between the types of decays also began to be examined: For example, gamma decay was almost always found to be associated with other types of decay, and occurred at about the same time, or afterwards. Gamma decay as a separate phenomenon, with its own half-life (now termed isomeric transition), was found in natural radioactivity to be a result of the gamma decay of excited metastable nuclear isomers, which were in turn created from other types of decay.

Although alpha, beta, and gamma radiations were most commonly found, other types of emission were eventually discovered. Shortly after the discovery of the positron in cosmic ray products, it was realized that the same process that operates in classical beta decay can also produce positrons (positron emission), along with neutrinos (classical beta decay produces antineutrinos). In a more common analogous process, called electron capture, some proton-rich nuclides were found to capture their own atomic electrons instead of emitting positrons, and subsequently these nuclides emit only a neutrino and a gamma ray from the excited nucleus (and often also Auger electrons and characteristic X-rays, as a result of the re-ordering of electrons to fill the place of the missing captured electron). These types of decay involve the nuclear capture of electrons or emission of electrons or positrons, and thus acts to move a nucleus toward the ratio of neutrons to protons that has the least energy for a given total number of nucleons. This consequently produces a more stable (lower energy) nucleus.

(A theoretical process of positron capture, analogous to electron capture, is possible in antimatter atoms, but has not been observed, as complex antimatter atoms beyond antihelium are not experimentally available. Such a decay would require antimatter atoms at least as complex as beryllium-7, which is the lightest known isotope of normal matter to undergo decay by electron capture.)

Shortly after the discovery of the neutron in 1932, Enrico Fermi realized that certain rare beta-decay reactions immediately yield neutrons as a decay particle (neutron emission). Isolated proton emission was eventually observed in some elements. It was also found that some heavy elements may undergo spontaneous fission into products that vary in composition. In a phenomenon called cluster decay, specific combinations of neutrons and protons other than alpha particles (helium nuclei) were found to be spontaneously emitted from atoms.

Other types of radioactive decay were found to emit previously-seen particles, but via different mechanisms. An example is internal conversion, which results in an initial electron emission, and then often further characteristic X-rays and Auger electrons emissions, although the internal conversion process involves neither beta nor gamma decay. A neutrino is not emitted, and none of the electron(s) and photon(s) emitted originate in the nucleus, even though the energy to emit all of them does originate there. Internal conversion decay, like isomeric transition gamma decay and neutron emission, involves the release of energy by an excited nuclide, without the transmutation of one element into another.

Rare events that involve a combination of two beta-decay type events happening simultaneously are known. Any decay process that does not violate the conservation of energy or momentum laws (and perhaps other particle conservation laws) is permitted to happen, although not all have been detected. An interesting example discussed in a final section, is bound state beta decay of rhenium-187. In this process, beta electron-decay of the parent nuclide is not accompanied by beta electron emission, because the beta particle has been captured into the K-shell of the emitting atom. An antineutrino is emitted, as in all negative beta decays.

Radionuclides can undergo a number of different reactions. These are summarized in the following table. A nucleus with mass number $A$ and atomic number $Z$ is represented as $(A, Z)$. The column "Daughter nucleus" indicates the difference between the new nucleus and the original nucleus. Thus, $(A - 1, Z)$ means that the mass number is one less than before, but the atomic number is the same as before.

If energy circumstances are favorable, a given radionuclide may undergo many competing types of decay, with some atoms decaying by one route, and others decaying by another. An example is copper-64, which has 29 protons, and 35 neutrons, which decays with a half-life of about 12.7 hours. This isotope has one unpaired proton and one unpaired neutron, so either the proton or the neutron can decay to the opposite particle. This particular nuclide (though not all nuclides in this situation) is almost equally likely to decay through positron emission (18%), or through electron capture (43%), as it does through electron emission (39%). The excited energy states resulting from these decays which fail to end in a ground energy state, also produce later internal conversion and gamma decay in almost 0.5% of the time.

More common in heavy nuclides is competition between alpha and beta decay. The daughter nuclides will then normally decay through beta or alpha, respectively, to end up in the same place.

| Mode of decay | Participating particles | Daughter nucleus |
|---|---|---|
| **Decays with emission of nucleons:** | | |
| Alpha decay | An alpha particle ($A = 4, Z = 2$) emitted from nucleus | $(A - 4, Z - 2)$ |
| Proton emission | A proton ejected from nucleus | $(A - 1, Z - 1)$ |
| Neutron emission | A neutron ejected from nucleus | $(A - 1, Z)$ |
| Double proton emission | Two protons ejected from nucleus simultaneously | $(A - 2, Z - 2)$ |
| Spontaneous fission | Nucleus disintegrates into two or more smaller nuclei and other particles | — |
| Cluster decay | Nucleus emits a specific type of smaller nucleus $(A_1, Z_1)$ which is larger than an alpha particle | $(A - A_1, Z - Z_1) + (A_1, Z_1)$ |
| **Different modes of beta decay:** | | |
| $\beta^-$ decay | A nucleus emits an electron and an electron antineutrino | $(A, Z + 1)$ |
| Positron emission ($\beta^+$ decay) | A nucleus emits a positron and an electron neutrino | $(A, Z - 1)$ |
| Electron capture | A nucleus captures an orbiting electron and emits a neutrino; the daughter nucleus is left in an excited unstable state | $(A, Z - 1)$ |
| Bound state beta decay | A free neutron or nucleus beta decays to electron and antineutrino, but the electron is not emitted, as it is captured into an empty K-shell; the daughter nucleus is left in an excited and unstable state. This process is a minority of free neutron decays (0.0004%) due to the low energy of hydrogen ionization, and is suppressed except in ionized atoms that have K-shell vacancies. | $(A, Z + 1)$ |
| Double beta decay | A nucleus emits two electrons and two antineutrinos | $(A, Z + 2)$ |
| Double electron capture | A nucleus absorbs two orbital electrons and emits two neutrinos – the daughter nucleus is left in an excited and unstable state | $(A, Z - 2)$ |

| | | |
|---|---|---|
| Electron capture with positron emission | A nucleus absorbs one orbital electron, emits one positron and two neutrinos | $(A, Z-2)$ |
| Double positron emission | A nucleus emits two positrons and two neutrinos | $(A, Z-2)$ |
| **Transitions between states of the same nucleus:** | | |
| Isomeric transition | Excited nucleus releases a high-energy photon (gamma ray) | $(A, Z)$ |
| Internal conversion | Excited nucleus transfers energy to an orbital electron, which is subsequently ejected from the atom | $(A, Z)$ |

Radioactive decay results in a reduction of summed rest mass, once the released energy (the *disintegration energy*) has escaped in some way. Although decay energy is sometimes defined as associated with the difference between the mass of the parent nuclide products and the mass of the decay products, this is true only of rest mass measurements, where some energy has been removed from the product system. This is true because the decay energy must always carry mass with it, wherever it appears according to the formula $E = mc^2$. The decay energy is initially released as the energy of emitted photons plus the kinetic energy of massive emitted particles (that is, particles that have rest mass). If these particles come to thermal equilibrium with their surroundings and photons are absorbed, then the decay energy is transformed to thermal energy, which retains its mass.

Decay energy therefore remains associated with a certain measure of mass of the decay system, called invariant mass, which does not change during the decay, even though the energy of decay is distributed among decay particles. The energy of photons, the kinetic energy of emitted particles, and, later, the thermal energy of the surrounding matter, all contribute to the invariant mass of the system. Thus, while the sum of the rest masses of the particles is not conserved in radioactive decay, the *system* mass and system invariant mass (and also the system total energy) is conserved throughout any decay process. This is a restatement of the equivalent laws of conservation of energy and conservation of mass.

## Radioactive Decay Rates

The *decay rate*, or *activity*, of a radioactive substance is characterized by:

Constant Quantities:

- The *half-life*—$t_{1/2}$, is the time taken for the activity of a given amount of a radioactive substance to decay to half of its initial value;
- The *decay constant*— $\lambda$, "lambda" the inverse of the mean lifetime, sometimes referred to as simply *decay rate*.
- The *mean lifetime*— $\tau$, "tau" the average lifetime (1/e life) of a radioactive particle before decay.

Although these are constants, they are associated with the statistical behavior of populations of atoms. In consequence, predictions using these constants are less accurate for minuscule samples of atoms.

In principle a half-life, a third-life, or even a (1/√2)-life, can be used in exactly the same way as half-life; but the mean life and half-life $t_{1/2}$ have been adopted as standard times associated with exponential decay.

Time-variable Quantities:

- *Total activity*— $A$, is the number of decays per unit time of a radioactive sample.
- *Number of particles*—$N$, is the total number of particles in the sample.
- *Specific activity*—$S_A$, number of decays per unit time per amount of substance of the sample at time set to zero ($t$ = 0). "Amount of substance" can be the mass, volume or moles of the initial sample.

These are related as follows:

$$t_{1/2} = \frac{\ln(2)}{\lambda} = \tau \ln(2)$$

$$A = -\frac{dN}{dt} = \lambda N$$

$$S_A a_0 = -\frac{dN}{dt}\Big|_{t=0} = \lambda N_0$$

where $N_0$ is the initial amount of active substance — substance that has the same percentage of unstable particles as when the substance was formed.

## Mathematics of Radioactive Decay

## Universal Law of Radioactive Decay

Radioactivity is one very frequently given example of exponential decay. The law describes the statistical behaviour of a large number of nuclides, rather than individual atoms. In the following formalism, the number of nuclides or the nuclide population $N$, is of course a discrete variable (a natural number)—but for any physical sample $N$ is so large that it can be treated as a continuous variable. Differential calculus is used to model the behaviour of nuclear decay.

The mathematics of radioactive decay depend on a key assumption that a nucleus of a radionuclide has no "memory" or way of translating its history into its present behavior. A nucleus does not "age" with the passage of time. Thus, the probability of its breaking down does not increase with time, but stays constant no matter how long the nucleus has existed. This constant probability may vary greatly between different types of nuclei, leading to the many different observed decay rates. However, whatever the probability is, it does not change. This is in marked contrast to complex objects which do show aging, such as automobiles and humans. These systems do have a chance of breakdown per unit of time, that increases from the moment they begin their existence.

## One-decay Process

Consider the case of a nuclide $A$ that decays into another $B$ by some process $A \rightarrow B$ (emission of other particles, like electron neutrinos νe and electrons $e^-$ as in beta decay, are irrelevant in what follows). The decay of an unstable nucleus is entirely random and it is impossible to predict when a particular atom will decay. However, it is equally likely to decay at any instant in time. Therefore, given a sample of a particular radioisotope, the number of decay events $-dN$ expected to occur in a small interval of time $dt$ is proportional to the number of atoms present $N$, that is

$$-\frac{dN}{dt} \propto N.$$

Particular radionuclides decay at different rates, so each has its own decay constant λ. The expected decay $-dN/N$ is proportional to an increment of time, $dt$:

$$-\frac{dN}{N} = \lambda dt$$

The negative sign indicates that $N$ decreases as time increases, as the decay events follow one after another. The solution to this first-order differential equation is the function:

$$N(t) = N_0 e^{-\lambda t} = N_0 e^{-t/\tau},$$

where $N_0$ is the value of $N$ at time $t = 0$.

We have for all time $t$:

$$N_A + N_B = N_{\text{total}} = N_{A0},$$

where $N_{\text{total}}$ is the constant number of particles throughout the decay process, which is equal to the initial number of $A$ nuclides since this is the initial substance.

If the number of non-decayed $A$ nuclei is:

$$N_A = N_{A0} e^{-\lambda t}$$

then the number of nuclei of $B$, i.e. the number of decayed $A$ nuclei, is

$$N_B = N_{A0} - N_A = N_{A0} - N_{A0} e^{-\lambda t} = N_{A0}\left(1 - e^{-\lambda t}\right).$$

The number of decays observed over a given interval obeys Poisson statistics. If the average number of decays is $\langle N \rangle$, the probability of a given number of decays $N$ is

$$P(N) = \frac{\langle N \rangle^N \exp(-\langle N \rangle)}{N!}.$$

## Chain-decay Processes

### Chain of Two Decays

Now consider the case of a chain of two decays: one nuclide $A$ decaying into another $B$ by one process, then $B$ decaying into another $C$ by a second process, i.e. $A \rightarrow B \rightarrow C$. The previous equation cannot be applied to the decay chain, but can be generalized as follows. Since $A$ decays into $B$, *then* $B$ decays into $C$, the activity of $A$ adds to the total number of $B$ nuclides in the present sample, *before* those $B$ nuclides decay and reduce the number of nuclides leading to the later sample. In other words, the number of second generation nuclei $B$ increases as a result of the first generation nuclei decay of $A$, and decreases as a result of its own decay into the third generation nuclei $C$. The sum of these two terms gives the law for a decay chain for two nuclides:

$$\frac{\mathrm{d}N_B}{\mathrm{d}t} = -\lambda_B N_B + \lambda_A N_A.$$

The rate of change of $N_B$, that is $\mathrm{d}N_B/\mathrm{d}t$, is related to the changes in the amounts of $A$ and $B$, $N_B$ can increase as $B$ is produced from $A$ and decrease as $B$ produces $C$.

Re-writing using the previous results:

$$\frac{\mathrm{d}N_B}{\mathrm{d}t} = -\lambda_B N_B + \lambda_A N_{A0} e^{-\lambda_A t}$$

The subscripts simply refer to the respective nuclides, i.e. $N_A$ is the number of nuclides of type $A$, $N_{A0}$ is the initial number of nuclides of type $A$, $\lambda_A$ is the decay constant for $A$ - and similarly for nuclide $B$. Solving this equation for $N_B$ gives:

$$N_B = \frac{N_{A0}\lambda_A}{\lambda_B - \lambda_A}\left(e^{-\lambda_A t} - e^{-\lambda_B t}\right).$$

In the case where $B$ is a stable nuclide ($\lambda_B = 0$), this equation reduces to the previous solution:

$$\lim_{\lambda_B \to 0}\left[\frac{N_{A0}\lambda_A}{\lambda_B - \lambda_A}\left(e^{-\lambda_A t} - e^{-\lambda_B t}\right)\right] = \frac{N_{A0}\lambda_A}{0 - \lambda_A}\left(e^{-\lambda_A t} - 1\right) = N_{A0}\left(1 - e^{-\lambda_A t}\right),$$

as shown above for one decay. The solution can be found by the integration factor method, where the integrating factor is $e^{\lambda_B t}$. This case is perhaps the most useful, since it can derive both the one-decay equation and the equation for multi-decay chains more directly.

### Chain of any Number of Decays

For the general case of any number of consecutive decays in a decay chain, i.e. $A_1 \rightarrow A_2 \cdots \rightarrow A_i \cdots \rightarrow A_D$, where $D$ is the number of decays and $i$ is a dummy index ($i = {}_{1, 2, 3, \ldots D}$), each nuclide population can be found in terms of the previous population. In this case $N_2 = 0$, $N_3 = 0, \ldots, N_D = 0$. Using the above result in a recursive form:

$$\frac{dN_j}{dt} = -\lambda_j N_j + \lambda_{j-1} N_{(j-1)0} e^{-\lambda_{j-1} t}.$$

The general solution to the recursive problem is given by *Bateman's equations*:

Bateman's equations

$$N_D = \frac{N_1(0)}{\lambda_D} \sum_{i=1}^{D} \lambda_i c_i e^{-\lambda_i t}$$

$$c_i = \prod_{j=1, i \neq j}^{D} \frac{\lambda_j}{\lambda_j - \lambda_i}$$

## Alternative Decay Modes

In all of the above examples, the initial nuclide decays into just one product. Consider the case of one initial nuclide that can decay into either of two products, that is $A \rightarrow B$ and $A \rightarrow C$ in parallel. For example, in a sample of potassium-40, 89.3% of the nuclei decay to calcium-40 and 10.7% to argon-40. We have for all time $t$:

$$N = N_A + N_B + N_C$$

which is constant, since the total number of nuclides remains constant. Differentiating with respect to time:

$$\frac{dN_A}{dt} = -\left(\frac{dN_B}{dt} + \frac{dN_C}{dt}\right)$$
$$-\lambda N_A = -N_A\left(\lambda_B + \lambda_C\right)$$

defining the *total decay constant* $\lambda$ in terms of the sum of *partial decay constants* $\lambda_B$ and $\lambda_C$:

$$\lambda = \lambda_B + \lambda_C.$$

Notice that

$$\frac{dN_A}{dt} < 0, \frac{dN_B}{dt} > 0, \frac{dN_C}{dt} > 0.$$

Solving this equation for $N_A$:

$$N_A = N_{A0} e^{-\lambda t}.$$

where $N_{A0}$ is the initial number of nuclide A. When measuring the production of one nuclide, one can only observe the total decay constant $\lambda$. The decay constants $\lambda_B$ and $\lambda_C$ determine the probability for the decay to result in products $B$ or $C$ as follows:

$$N_B = \frac{\lambda_B}{\lambda} N_{A0}\left(1 - e^{-\lambda t}\right),$$

$$N_C = \frac{\lambda_C}{\lambda} N_{A0}\left(1 - e^{-\lambda t}\right).$$

because the fraction $\lambda_B/\lambda$ of nuclei decay into $B$ while the fraction $\lambda_C/\lambda$ of nuclei decay into $C$.

## Corollaries of the Decay Laws

The above equations can also be written using quantities related to the number of nuclide particles $N$ in a sample;

- The activity: $A = \lambda N$.
- The amount of substance: $n = N/L$.
- The mass: $M = A_r n = A_r N/L$.

where $L = 6.022\times10^{23}$ is Avogadro's constant, $A_r$ is the relative atomic mass number, and the amount of the substance is in moles.

## Decay Timing: Definitions and Relations

### Time Constant and Mean-life

For the one-decay solution $A \rightarrow B$:

$$N = N_0 e^{-\lambda t} = N_0 e^{-t/\tau},$$

the equation indicates that the decay constant $\lambda$ has units of $t^{-1}$, and can thus also be represented as $1/\tau$, where $\tau$ is a characteristic time of the process called the *time constant*.

In a radioactive decay process, this time constant is also the mean lifetime for decaying atoms. Each atom "lives" for a finite amount of time before it decays, and it may be shown that this mean lifetime is the arithmetic mean of all the atoms' lifetimes, and that it is $\tau$, which again is related to the decay constant as follows:

$$\tau = \frac{1}{\lambda}.$$

This form is also true for two-decay processes simultaneously $A \rightarrow B + C$, inserting the equivalent values of decay constants

$$\lambda = \lambda_B + \lambda_C$$

into the decay solution leads to:

$$\frac{1}{\tau} = \lambda = \lambda_B + \lambda_C = \frac{1}{\tau_B} + \frac{1}{\tau_C}$$

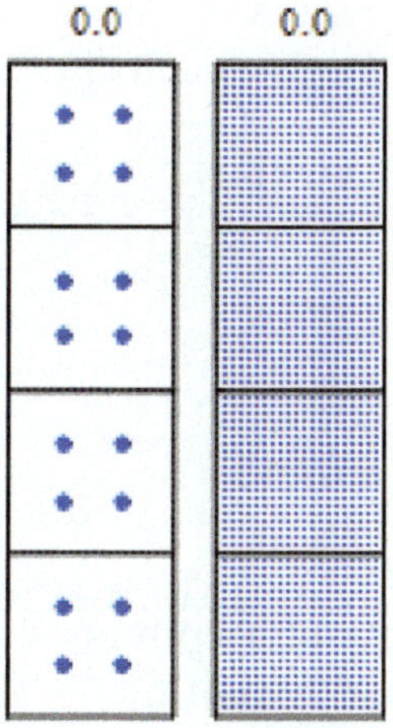


Simulation of many identical atoms undergoing radioactive decay, starting with either 4 atoms (left) or 400 (right). The number at the top indicates how many half-lives have elapsed.

## Half-life

A more commonly used parameter is the half-life. Given a sample of a particular radionuclide, the half-life is the time taken for half the radionuclide's atoms to decay. For the case of one-decay nuclear reactions:

$$N = N_0 e^{-\lambda t} = N_0 e^{-t/\tau},$$

the half-life is related to the decay constant as follows: set $N = N_o/2$ and $t = T_{1/2}$ to obtain

$$t_{1/2} = \frac{\ln 2}{\lambda} = \tau \ln 2.$$

This relationship between the half-life and the decay constant shows that highly radioactive substances are quickly spent, while those that radiate weakly endure longer. Half-lives of known radionuclides vary widely, from more than $10^{19}$ years, such as for the very nearly stable nuclide $^{209}Bi$, to $10^{-23}$ seconds for highly unstable ones.

The factor of ln(2) in the above relations results from the fact that the concept of "half-life" is merely a way of selecting a different base other than the natural base *e* for the lifetime expression. The time constant $\tau$ is the $e^{-1}$ -life, the time until only $1/e$ remains, about 36.8%, rather than the 50% in the half-life of a radionuclide. Thus, $\tau$ is longer than $t_{1/2}$. The following equation can be shown to be valid:

$$N(t) = N_0 e^{-t/\tau} = N_0 2^{-t/t_{1/2}}.$$

Since radioactive decay is exponential with a constant probability, each process could as easily be described with a different constant time period that (for example) gave its "(1/3)-life" (how long until only 1/3 is left) or "(1/10)-life" (a time period until only 10% is left), and so on. Thus, the choice of $\tau$ and $t_{1/2}$ for marker-times, are only for convenience, and from convention. They reflect a fundamental principle only in so much as they show that the *same proportion* of a given radioactive substance will decay, during any time-period that one chooses.

Mathematically, the $n^{th}$ life for the above situation would be found in the same way as above—by setting $N = N_o/n$, $t = T_{1/n}$ and substituting into the decay solution to obtain

$$t_{1/n} = \frac{\ln n}{\lambda} = \tau \ln n.$$

## Example

A sample of $^{14}C$ has a half-life of 5,730 years and a decay rate of 14 disintegration per minute (dpm) per gram of natural carbon.

If an artifact is found to have radioactivity of 4 dpm per gram of its present C, we can find the approximate age of the object using the above equation:

$$N = N_0 e^{-t/\tau},$$

where: $\frac{N}{N_0} = 4/14 \approx 0.286,$

$$\tau = \frac{T_{1/2}}{\ln 2} \approx 8267 \text{ years},$$

$$t = -\tau \ln \frac{N}{N_0} \approx 10356 \text{ years}.$$

## Changing Decay Rates

The radioactive decay modes of electron capture and internal conversion are known to be slightly sensitive to chemical and environmental effects that change the electronic structure of the atom, which in turn affects the presence of 1s and 2s electrons that participate in the decay process. A small number of mostly light nuclides are affected. For example, chemical bonds can affect the rate of electron capture to a small degree (in general, less than 1%) depending on the proximity of electrons to the nucleus. In $^7Be$, a difference of 0.9% has been observed between half-lives in metallic and insulating environments. This relatively large effect is because beryllium is a small atom whose valence electrons are in 2s atomic orbitals, which are subject to electron capture in $^7Be$ because (like all s atomic orbitals in all atoms) they naturally penetrate into the nucleus.

In 1992, Jung et al. of the Darmstadt Heavy-Ion Research group observed an accelerated $\beta^-$ decay of $^{163}Dy^{66+}$. Although neutral $^{163}Dy$ is a stable isotope, the fully ionized $^{163}Dy^{66+}$ undergoes $\beta^-$ decay into the K and L shells to $^{163}Ho^{66+}$ with a half-life of 47 days.

Rhenium-187 is another spectacular example. $^{187}Re$ normally beta decays to $^{187}Os$ with a half-life of $41.6 \times 10^9$ years, but studies using fully ionised $^{187}Re$ atoms (bare nuclei) have found that this can decrease to only 33 years. This is attributed to "bound-state $\beta^-$ decay" of the fully ionised atom – the electron is emitted into the "K-shell" (1s atomic orbital), which cannot occur for neutral atoms in which all low-lying bound states are occupied.

A number of experiments have found that decay rates of other modes of artificial and naturally occurring radioisotopes are, to a high degree of precision, unaffected by external conditions such as temperature, pressure, the chemical environment, and electric, magnetic, or gravitational fields. Comparison of laboratory experiments over the last century, studies of the Oklo natural nuclear reactor (which exemplified the effects of thermal neutrons on nuclear decay), and astrophysical observations of the luminosity decays of distant supernovae (which occurred far away so the light has taken a great deal of time to reach us), for example, strongly indicate that unperturbed decay rates have been constant (at least to within the limitations of small experimental errors) as a function of time as well.

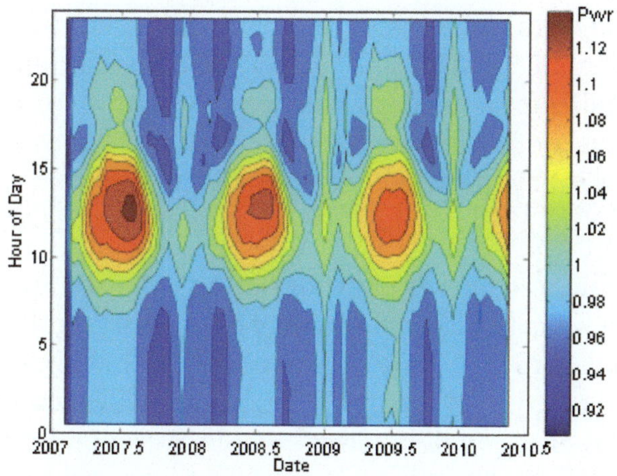


Decay rate of radon-222 as a function of date and time of day. The color-bar gives the power of the observed signal and represents ~4% seasonal decay rate variation.

Recent results suggest the possibility that decay rates might have a weak dependence on environmental factors. It has been suggested that measurements of decay rates of silicon-32, manganese-54, and radium-226 exhibit small seasonal variations (of the order of 0.1%), while the decay of radon-222 is reported to exhibit large 4% peak-to-peak seasonal variations, proposed to be related to either solar flare activity or the distance from the Sun. However, such measurements are highly susceptible to systematic errors, and a subsequent paper has found no evidence for such correlations in seven other isotopes ($^{22}$Na, $^{44}$Ti, $^{108}$Ag, $^{121}$Sn, $^{133}$Ba, $^{241}$Am, $^{238}$Pu), and sets upper limits on the size of any such effects.

## GSI Anomaly

An unexpected series of experimental results for the rate of decay of heavy highly charged radioactive ions circulating in a storage ring has provoked theoretical activity in an effort to find a convincing explanation. The rates of weak decay of two radioactive species with half lives of about 40 s and 200 s are found to have a significant oscillatory modulation, with a period of about 7 s. The observed phenomenon is known as the GSI anomaly, as the storage ring is a facility at the GSI Helmholtz Centre for Heavy Ion Research in Darmstadt Germany. As the decay process produces an electron neutrino, some of the proposed explanations for the observed rate oscillation invoke neutrino properties. Initial ideas related to flavour oscillation met with skepticism. A more recent proposal involves mass differences between neutrino mass eigenstates.

## Theoretical Basis of Decay Phenomena

The neutrons and protons that constitute nuclei, as well as other particles that approach close enough to them, are governed by several interactions. The strong nuclear force, not observed at the familiar macroscopic scale, is the most powerful force over subatomic distances. The electrostatic force is almost always significant, and, in the case of beta decay, the weak nuclear force is also involved.

The interplay of these forces produces a number of different phenomena in which energy may be released by rearrangement of particles in the nucleus, or else the change of one type of particle into others. These rearrangements and transformations may be hindered energetically, so that they do not occur immediately. In certain cases, random quantum vacuum fluctuations are theorized to promote relaxation to a lower energy state (the "decay") in a phenomenon known as quantum tunneling. Radioactive decay half-life of nuclides has been measured over timescales of 55 orders of magnitude, from 2.3 x $10^{-23}$ seconds (for hydrogen-7) to 6.9 x $10^{31}$ seconds (for tellurium-128). The limits of these timescales are set by the sensitivity of instrumentation only, and there are no known natural limits to how brief or long a decay half life for radioactive decay of a radionuclide may be.

The decay process, like all hindered energy transformations, may be analogized by a snowfield on a mountain. While friction between the ice crystals may be supporting the snow's weight, the system is inherently unstable with regard to a state of lower potential energy. A disturbance would thus facilitate the path to a state of greater entropy: The system will move towards the ground state, producing heat, and the total energy will be distributable over a larger number of quantum states thus resulting in an avalanche . The *total* energy does not change in this process, but, because of the second law of thermodynamics, avalanches have only been observed in one direction and that is toward the "ground state" — the state with the largest number of ways in which the available energy could be distributed.

Such a collapse (a *decay event*) requires a specific activation energy. For a snow avalanche, this energy comes as a disturbance from outside the system, although such disturbances can be arbitrarily small. In the case of an excited atomic nucleus, the arbitrarily small disturbance comes from quantum vacuum fluctuations. A radioactive nucleus (or any excited system in quantum mechanics) is unstable, and can, thus, *spontaneously* stabilize to a less-excited system. The resulting transformation alters the structure of the nucleus and results in the emission of either a photon or a high-velocity particle that has mass (such as an electron, alpha particle, or other type).

## Occurrence and Applications

According to the Big Bang theory, stable isotopes of the lightest five elements (H, He, and traces of Li, Be, and B) were produced very shortly after the emergence of the universe, in a process called Big Bang nucleosynthesis. These lightest stable nuclides (including deuterium) survive to today, but any radioactive isotopes of the light elements produced in the Big Bang (such as tritium) have long since decayed. Isotopes of elements heavier than boron were not produced at all in the Big Bang, and these first five elements do not have any long-lived radioisotopes. Thus, all radioactive nuclei are, therefore, relatively young with respect to the birth of the universe, having formed later in various other types of nucleosynthesis in stars (in particular, supernovae), and also during ongoing interactions between stable isotopes and energetic particles. For example, carbon-14, a radioactive nuclide with a half-life of only 5,730 years, is constantly produced in Earth's upper atmosphere due to interactions between cosmic rays and nitrogen.

Nuclides that are produced by radioactive decay are called radiogenic nuclides, whether they themselves are stable or not. There exist stable radiogenic nuclides that were formed from short-lived extinct radionuclides in the early solar system. The extra presence of these stable radiogenic nuclides (such as Xe-129 from primordial I-129) against the background of primordial stable nuclides can be inferred by various means.

Radioactive decay has been put to use in the technique of radioisotopic labeling, which is used to track the passage of a chemical substance through a complex system (such as a living organism). A sample of the substance is synthesized with a high concentration of unstable atoms. The presence of the substance in one or another part of the system is determined by detecting the locations of decay events.

On the premise that radioactive decay is truly random (rather than merely chaotic), it has been used in hardware random-number generators. Because the process is not thought to vary significantly in mechanism over time, it is also a valuable tool in estimating the absolute ages of certain materials. For geological materials, the radioisotopes and some of their decay products become trapped when a rock solidifies, and can then later be used (subject to many well-known qualifications) to estimate the date of the solidification. These include checking the results of several simultaneous processes and their products against each other, within the same sample. In a similar fashion, and also subject to qualification, the rate of formation of carbon-14 in various eras, the date of formation of organic matter within a certain period related to the isotope's half-life may be estimated, because the carbon-14 becomes trapped when the organic matter grows and incorporates the new carbon-14 from the air. Thereafter, the amount of carbon-14 in organic matter decreases according to decay processes that may also be independently cross-checked by other means (such as checking the carbon-14 in individual tree rings, for example).

## Szilard–Chalmers Effect

The *Szilard–Chalmers effect* is defined as the breaking of a chemical bond between an atom and the molecule that the atom is part of, as a result of a nuclear reaction of the atom. The effect can be used to separate isotopes by chemical means. The discovery of this effect is due to L. Szilárd and T.A. Chalmers.

## Origins of Radioactive Nuclides

Radioactive primordial nuclides found in the Earth are residues from ancient supernova explosions that occurred before the formation of the solar system. They are the fraction of radionuclides that survived from that time, through the formation of the primordial solar nebula, through planet accretion, and up to the present time. The naturally occurring short-lived radiogenic radionuclides found in today's rocks, are the daughters of those radioactive primordial nuclides. Another minor source of naturally occurring radioactive nuclides are cosmogenic nuclides, that are formed by cosmic ray bombardment of material in the Earth's atmosphere or crust. The decay of the radionuclides in rocks of the Earth's mantle and crust contribute significantly to Earth's internal heat budget.

## Decay Chains and Multiple Modes

The daughter nuclide of a decay event may also be unstable (radioactive). In this case, it too will decay, producing radiation. The resulting second daughter nuclide may also be radioactive. This can lead to a sequence of several decay events called a *decay chain*. Eventually, a stable nuclide is produced.

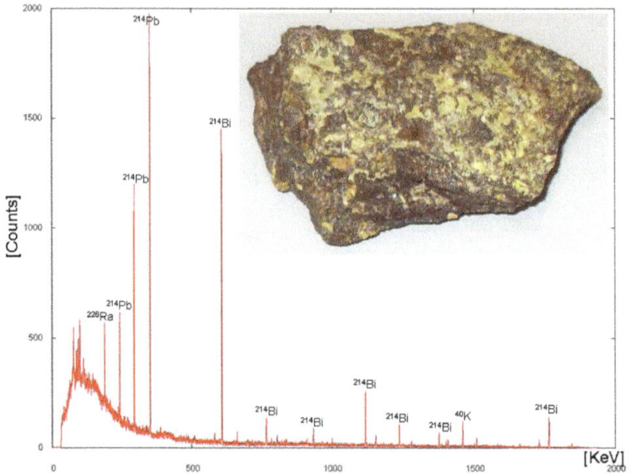


Gamma-ray energy spectrum of uranium ore (inset). Gamma-rays are emitted by decaying nuclides, and the gamma-ray energy can be used to characterize the decay (which nuclide is decaying to which). Here, using the gamma-ray spectrum, several nuclides that are typical of the decay chain of $^{238}U$ have been identified: $^{226}Ra$, $^{214}Pb$, $^{214}Bi$.

An example is the natural decay chain of $^{238}U$:

- Uranium-238 decays, through alpha-emission, with a half-life of 4.5 billion years to thorium-234
- which decays, through beta-emission, with a half-life of 24 days to protactinium-234
- which decays, through beta-emission, with a half-life of 1.2 minutes to uranium-234
- which decays, through alpha-emission, with a half-life of 240 thousand years to thorium-230
- which decays, through alpha-emission, with a half-life of 77 thousand years to radium-226
- which decays, through alpha-emission, with a half-life of 1.6 thousand years to radon-222
- which decays, through alpha-emission, with a half-life of 3.8 days to polonium-218
- which decays, through alpha-emission, with a half-life of 3.1 minutes to lead-214
- which decays, through beta-emission, with a half-life of 27 minutes to bismuth-214
- which decays, through beta-emission, with a half-life of 20 minutes to polonium-214
- which decays, through alpha-emission, with a half-life of 160 microseconds to lead-210
- which decays, through beta-emission, with a half-life of 22 years to bismuth-210
- which decays, through beta-emission, with a half-life of 5 days to polonium-210
- which decays, through alpha-emission, with a half-life of 140 days to lead-206, which is a stable nuclide.

Some radionuclides may have several different paths of decay. For example, approximately 36% of bismuth-212 decays, through alpha-emission, to thallium-208 while approximately 64% of bismuth-212 decays, through beta-emission, to polonium-212. Both thallium-208 and polonium-212 are radioactive daughter products of bismuth-212, and both decay directly to stable lead-208.

## Associated Hazard Warning Signs

The trefoil symbol used to indicate ionising radiation.

2007 ISO radioactivity danger symbol intended for IAEA Category 1, 2 and 3 sources defined as dangerous sources capable of death or serious injury.

The dangerous goods transport classification sign for radioactive materials

## Law of Radioactive Decay

When a nucleus disintegrates by emitting charged particles (α, β) or a γ- ray photon, or by capturing electron from the atomic shell, the process is called radioactive decay. This decay is spontaneous. The rate of change of nuclei in a radioactive sample is given by law of radioactive decay. Let us take a radioactive sample containing No nuclei at time t = 0, i.e., at the beginning. We wish to calculate the number N of these nuclei left after time t. The number of nuclei of a radioactive sample disintegrating per second is called the activity of that sample.

$\frac{dN}{dt}$ = Rate of decrease of nuclei with time

= Activity at time t.

Experimentally, it has been observed that the activity at any instant of time t is directly proportional to the number N of parent type nuclei present at that time.

$$-\frac{dN}{dt}\alpha\ N$$

Or

$$-\frac{dN}{dt}=\lambda N$$

where $\lambda > 0$, is the proportionality constant known as decay constant.

The negative sign indicates that N decreases as t increases. By integrating the equation we get,

$$\int_{N_0}^{N}\frac{dN}{N}=-\int_{0}^{t}\lambda dt$$

or

$$N=N_0 e^{-\lambda t}$$

Thus we see that the law of radioactive decay is exponential in nature. Number of radioactive nuclei versus time is plotted in the figure.

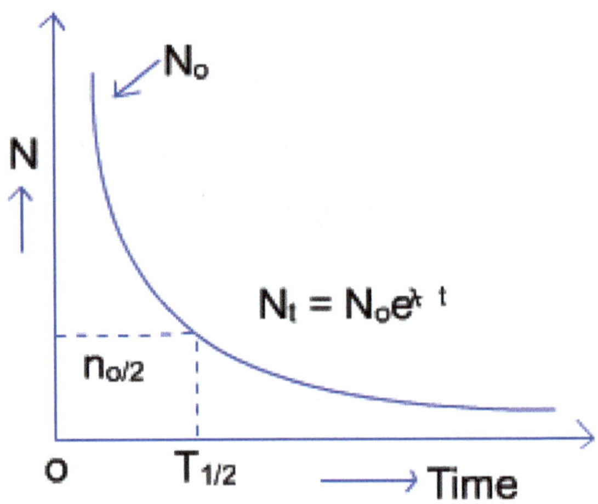


Number of radioactive nuclei with time showing the exponential nature of radioactive decay.

## Theory of Successive Transformation and Radioactive Equilibrium

Any radioactive atom decays to its daughter atom following its characteristics decay constant. The daughter atom, if radioactive, decays to third generation and in this manner a chain of decay propagates, which eventually stops by the formation of a stable nucleus. Let us consider a radioactive nuclide, symbolized by a subscript p (Parent nucleus) to decay into another radioactive nuclide (subscript d: daughter nucleus); which again decays to a stable end product. Let $\lambda_p$, $\lambda_d$ be the decay constants of nuclides p and d respectively, and $N_p$, $N_d$ and N3 be the number of atoms of the three kinds at any instant. The second nuclide would be formed at the rate of $\lambda_p N_p$ by the decay of the parent atom, and would decay at the rate of $\lambda_d N_d$. In the same manner, the atoms of the end product will appear at the rate λdNd by the decay of nuclide, but being stable will not decay. Therefore,

$$\frac{dN_p}{dt} = -\lambda_p N_p$$

$$\frac{dN_d}{dt} = \lambda_p N_p - \lambda_d N_d$$

$$\frac{dN_3}{dt} = \lambda_d N_d$$

Solving the equation, we have

$$N_p^t = N_p^0 e^{-\lambda t}$$

where, $N_p^0$ is the number of atoms of nuclide p at t = o.

we have:

$$\frac{dN_d}{dt} = \lambda_p N_p^0 e^{-\lambda t} - \lambda_d N_d$$

$$or \quad \frac{dN_d}{dt} + \lambda_d N_d = \lambda_p N_p^0 e^{-\lambda t}$$

Multiplying both sides by $e^{\lambda_d t}$ and then integrating, we get

$$N_d e^{\lambda t} = \frac{\lambda_p}{\lambda_d - \lambda_p} N^0 e^{(\lambda - \lambda)t} + A$$

where, A is the constant of integration.

Now, at t =o,$N_d$ = $N^o{}_d$ = a constant. So, from the equation,

$$A = N_d^0 - \frac{\lambda_p}{\lambda_d - \lambda_p} N_p^0$$

$$\therefore N_d^t = \frac{\lambda_p}{\lambda_d - \lambda_p} N_p^0 (e^{\lambda pt} - e^{-\lambda t}) + N_d^0 e^{-\lambda t}$$

From above equations,

$$N_3 = \left(\frac{\lambda_p}{\lambda_d - \lambda_p} N_p^0 - N_d^0\right) e^{-\lambda t} - \frac{\lambda_p}{\lambda_d - \lambda_p} N_p^0 e^{-\lambda t} + B$$

where, B is the integration constant. Now at t =0,$N_3$ = $N^o_3$, so that we have from the equation,

$$B = N_3^0 + N_d^0 + N_p^0$$

In practical cases, as only the first nuclide is present at t = 0, we can put $N^o_d$ = $N^o_3$ and from the equation, we get

$$N_d^t = \frac{\lambda_p}{\lambda_d - \lambda_p} N_p^0 (e^{-\lambda t} - e^{-\lambda t})$$

$$N_3^t = N_p^0 \left(1 + \frac{\lambda_p}{\lambda_d - \lambda_p} e^{-\lambda t} - \frac{\lambda_p}{\lambda_d - \lambda_p} e^{-\lambda t}\right)$$

The decay of the parent nuclide and the growth and decay of the daughter nuclide are shown in the figure.

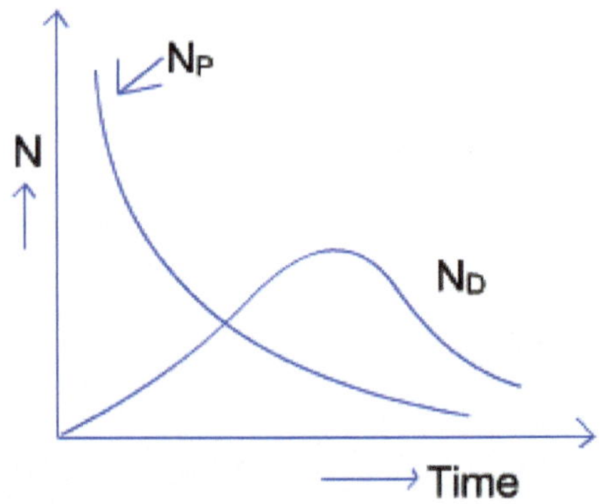


Decay of the parent nuclide and growth and decay of the first daughter nuclide with time

Example m4.2: Show that if $N^o_p$ be the number of parent atoms initially and $\gamma_p$ and $\gamma_d$ the disintegration constants of parent and daughter respectively, the time at which the number of daughter atoms is maximum is given by

$$t_m = \frac{In(\lambda_d / \lambda_p)}{\lambda_d - \lambda_p}$$

Solution:

From the relation m4.22, we get, $N_d^t = \frac{(\lambda_d / \lambda_p)}{\lambda_d - \lambda_p} N_p^0 (e^{-\lambda t} - e^{-\lambda t})$ , where, $N_d^t$ = No. of daughter atoms at any time t. For $N_d^t$ to be maximum, the condition is $\frac{dN_d^t}{dt} = 0$

$\Rightarrow -\lambda_p e^{-\lambda t} + \lambda_d e^{-\lambda t}$ = 0. Denoting t as tm corresponding to maxima, we can write

$$t_m = \frac{In\left(\frac{\lambda_d}{\lambda_p}\right)}{\lambda_d - \lambda_p}$$

## Radioactive Equilibrium

When we consider a radioactive series, it is important to understand the equilibrium conditions of all the nuclei in the respective chain. Applying the mathematical condition of equilibrium, i.e., time derivative of any quantity is equal to zero, we can write from

$$\frac{dN_1}{dt} = -\lambda_1 N_1 = 0$$
$$\lambda_1 N_1 = \lambda_2 N_2$$
$$\lambda_2 N_2 = \lambda_3 N_3$$
$$\lambda_{n-1} N_{n-1} = \lambda_n N_n$$

Where 1, 2, - n and represents parent, daughter and so on upto n<sup>th</sup> nucleus. It is clear from the equation, that the first equation is not valid as it implies $\lambda_1$ to be zero. Therefore, the mathematical condition of equilibrium cannot be applied here rigorously. However, some conditions can be set in to attain a state very close to equilibrium. Depending on the conditions, two equilibrium states can be defined:

## Secular Equilibrium ($\lambda_1$<< other $\lambda$'s)

In this case the first equation can be considered to certain approximation and hence we can write,

$$\lambda_1 N_1 = \lambda_2 N_2 = ..... = \lambda_{n-1} N_{n-1} = \lambda_n N_n$$

or

$$\frac{N_1}{(T_{1/2})_1} = \frac{N_2}{(T_{1/2})_2} = ----\frac{N_n}{(T_{1/2})_n}$$

As in this case $\lambda_1 \approx 0$ and also $\lambda_1 << \lambda_2, e^{-\lambda t} \approx 0$ Therefore, from the equation:

$$N_1^t \approx N_1^0$$

$$\therefore N_2^t \approx \frac{\lambda_1}{\lambda_2} N_1^0 (1 - e^{-\lambda t})$$

After a long time $e^{-\lambda t}{-}0$ , and from the equation, it can be written as

$$\lambda_2 N_2 \approx \lambda_1 N_1^0$$

This is known as secular equilibrium.

### Transient Equilibrium ($\lambda_1 < \lambda_2$ but $(T_{1/2})_1$ is not Very Long)

In this case $\lambda_1 \sim 0$ does not hold good. When t becomes large, $e^{-\lambda_2 t} << e^{-\lambda_1 t}$, Therefore from the equation:

$$N_2^t \approx \frac{\lambda_1}{\lambda_2 - \lambda_1} N_1^0 e^{-\lambda t}$$

This indicates that daughter decays with the same half life of the parent. This is basically the condition of transient equilibrium. From equations m4.1 and m4.28, we can write

$$\frac{N_2}{N_1} = \frac{\lambda_1}{\lambda_2 - \lambda_1}$$

Example m4.3: Calculate the amount of Ra-226 in secular equilibrium with 5 kg of pure U-238, given the half life of $^{226}Ra$ and $^{238}U$ as 1620 years and $4.5 \times 10^9$ years respectively.

Solution:

Let x gm be the required amount of $^{226}Ra$ in secular equilibrium with 5 kg of pure $^{238}U$. Now,

5 kg of pure U-238 = $\frac{5 \times 6.02 \times 10^{23}}{238} \times 10^3$ atoms of U

x gm of Ra-226 = $\frac{6.02 \times 10^{23}}{226} \times x$ atoms of Ra

From the condition of secular equilibrium

$$\frac{N_1}{(T_{1/2})_1} = \frac{N_2}{(T_{1/2})_2}$$

x = 1.7 mg.

## Nuclear Radiation with Biological Specimen

Nuclear radiation though hazardous to living organisms could be of immense use in medicine both for diagnostic and therapy. Anyone handling radioactive material or working in a radiation environment should have at least elementary ideas concerning the effects of exposure to radiation, the permissible limits and safety precautions. In order to use nuclear radiation in medicine, knowledge of radiation doses and limits is must.

## Dosimetric Units

The amount of nuclear radiation that is absorbed by anybody can be determined based on two processes, the ionization produced by the radiation and the energy deposited by it. The unit, which was used initially, is Roentgen, which is a measure of the exposure and is defined as

1 Roentgen (R) = the quantity of x-rays producing an ionization of 1 esu/$cm^3$

= ($2.58 \times 10^{-4}$ Coul/kg) in air at STP

This definition refers specifically to x-rays and γ-rays in air and can be measured easily with the help of a gas ionization chamber. However, it is inconvenient to measure in case of living tissue or some other material. A more relevant quantity to measure radiation dose is Gray (Gy), which is defined as

1 Gray (Gy) = 1 Joule/kg

This is also known as absorbed dose. It has been studied that the effect produced by any nuclear radiation on biological samples depends strongly on the type and energy of the incident radiation. An absorbed dose of α-particles produces more damage than an equal dose of protons. The difference lies in the linear energy transfer (LET) of the different particles, i.e. the energy locally deposited per unit path length of any specimen under study. Therefore, the radiation, which produces more ionization, causes higher damage. In order to account this effect, a radiation weighting factor, wR is associated with each radiation, the value of which is listed below in the table:

Radiation weighting factors

| Radiation type and energy | Radiation weighting factor ($W_R$) |
|---|---|
| Photons, all energies | 1 |
| Electrons and muons, all energies* | 1 |
| Neutrons<br>< 10 keV<br>10 keV to 100 keV<br>> 100 keV to 20 MeV<br>2 MeV to 20 MeV<br>> 20 MeV | <br>5<br>10<br>20<br>10<br>5 |
| Protons, other than recoil protons > 2MeV | 5 |
| α-particles, fission fragments, heavy nuclei | 20 |

If $D_R$ is the average absorbed dose received by any organ exposed to radiation, then equivalent dose is to be considered, which is defined as

$$H_r = \sum_R w_R D_{T,R}$$

where, $D_{T,R}$ is the average absorbed dose received by organ T from the radiation type R.

The damage produced by any radiation of any energy is different for different tissues and organs. To obtain a normalized measure of the biological effect suffered by a tissue or organ due to irradiation, tissue-weighting factor ($w_T$) is to be considered, which is then multiplied with equivalent dose to get the effective dose absorbed by any tissue. The tissue-weighting factor for different organs is tabulated below:

**TABLE m4.2 Tissue weighting factore[4]**

| Tissue or organ | Tissue weighting factor, $w_T$ |
|---|---|
| Bone marrow | 0.2 |
| Colon | 0.12 |
| Lung | 0.12 |
| Stomach | 0.12 |
| Bladder | 0.12 |
| Breast | 0.05 |
| Liver | 0.05 |
| Skin | 0.01 |
| Bone surface | 0.01 |

It is to be noted that the tissue-weighting factor is totally independent of the radiation type and energy, as the radiation-weighting factor is independent of tissue type. The effective dose is defined as

where, sum is over different tissues and organs exposed. The unit of equivalent dose is the Sievert (Sv), which has the same dimension as the Gray (J/kg). However, this dose is normalized by the radiobilogical effectiveness (RBE), so that 1 Sv of α-particles produces approximately same effect as 1 Sv of γ-rays.

Let us now discuss the the sequence of events on interaction of radiation with biological specimen. This can be divided in three phases

I. Physical phase

It is the first event for radiation damage of any biological sample. The basic physical phenomena involved in this phase are ionization, excitation absorption etc.

II. Chemical phase

Absorption of radiation producing ions, electrons and excited atoms induce chemical changes in the cell. The chemical changes occur by two ways:

(i) direct action, where molecules directly absorbs energy from radiation and

(ii) indirect action, where molecules absorb energy from other molecules, which acts as an intermediary for the transfer of energy. The intermediary in the cell is H2O, which by ionization, produces highly reactive ions and free radicals like $H^+$, $OH^-$ and $H_2O^*$ and their products like $H_2O_2$, which act as powerful oxidizing agents, react with important biomolecules like DNA,RNA and enzymes for alteration of their function. The chemical phase occurs within milliseconds.

III. Biological phase

The important biomolecules are intimately related to the cellular function. If they are damaged, the cellular function will be seriously hampered or there may be cellular death. Time involved in this process is from seconds to hours.

**TABLE m4.3 Changes in some important biomolecules by radiation[5]**

| Biomolecules | Changes due to radiation | Damaging effect in the cell |
|---|---|---|
| Enzymes | Changes in tertiary structure due to disruption of chemical bonds | Enzyme inhibition causing changes in metabolic reaction |
| DNA | Disruption of linear arrangement of bases | 1. Inhibition of DNA synthesis<br>2. Synthesis of defective DNA<br>3. Synthesis of defective protein |
| Structural lipids in the cell membrane | Disruption of molecular bonds | Increased permeability of the cell membrane to $Na^+$ and $K^+$ etc. |

Some important changes that occur in the cells due to radiation are:

1. Aberration (breakage) of chromosomes
2. Onset of mitosis is delayed followed by normal mitosis
3. Death of cells after few subdivisions
4. Death of cells without division
5. Transformation to malignancy.

## Radioisotopes and Their Use in Medical Diagnostics

Applications of nuclear radiation in diagnosis are of paramount importance in recent times. With the advancement of nuclear electronics and detector technology, nuclear radiation can be used to image different parts of the body with high sensitivity and resolution.

Let us first see examples of some radioisotopes and their important applications in biology.

### Radioactive Isotopes

Major number of natural isotopes is not radioactive, but they can be made radioactive artificially by bombardment of α-particles in a cyclotron or uranium pile. These nuclei again disintegrate by the emission of other particles. The half-lives of these isotopes may vary from few hours to several years. Some biologically important radioisotopes1, their mass number, type of radiation emitted, and half-lives are given in the table.

**Some biologically important radioisotopes**

| Element | Mass Number | Type of radiation | Half-life |
|---|---|---|---|
| Potassium | 42 | $\beta^-$, $\gamma$ | 12.5 hours |
| Sodium | 24 | $\beta^-$, $\gamma$ | 15 hours |
| Iodine | 131 | $\beta^-$, $\gamma$ | 8.1 days |
| Phosphorous | 32 | $\beta^-$, $\gamma$ | 14.3 days |
| Iron | 59 | $\beta^-$ | 45.1 days |
| Cobalt | 60 | $\beta^-$, $\gamma$ | 5.2 years |
| Carbon | 14 | $\beta^-$ | 5,570 years |

The radioisotope produced in the nuclear reactor is basically by the bombardment of neutrons and in cyclotrons they are produced by α-particle bombardment. The radioisotopes produced in cyclotron are mostly having shorter half-lives as compared to the radioisotope produced in the nuclear reactor. There are many applications of radioisotopes in medicine and biology. The following are the radinuclides having immense applications in nuclear medicine:

1. Iodine-123 and Iodine-125 are used in medicine as tracers for imaging and evaluating the function of the thyroid.
2. Iodine-131 is used in medicine for treatment of thyroid cancer and Grave's disease.
3. Cobalt-60 is largely used in radiotherapy.
4. Strontium-89 and Samarium-153 are used for the relief of cancer-induced bone pain.
5. Boron-10 is used in Boron neutron capture therapy a recent technique in cancer treatment.
6. Thallium-201 and Technetium-99 are used to detect the coronary artery disease and myocardial imaging.
7. Phosphorous-32 is used to control the growth of excess red blood corpuscles in bone marrow.
8. Chromium-51 is used to label red blood cells and quantify gastro-intestinal protein loss.
9. Copper-64 is used to study genetic diseases affecting copper metabolism, such as Wilson's and Menke's diseases.
10. Xenon-133 is used for pulmonary (lung) ventilation studies.

There are many other applications of radioisotopes as nuclear medicine. Radioisotopes are also used in Biochemical applications. They are under regular use to label the molecules of biological samples in vitro (out of the body). Many methods have been devised to determine the constituents of blood, serum, urine, hormones, antigens and many drugs by means of associated radioisotopes.

Now I am going to discuss two important techniques routinely used in modern medical science for imaging biological samples, organs and even the whole human body. The first imaging technique discussed here is Positron emission tomography (PET) where positron emitting radioisotope is used as a source and the second is magnetic resonance imaging (MRI), where not radioisotope is required. In MRI the magnetic moment of protons inside the body is used to get the information about any organ, pathology, fluid flow etc.

Positron Emission Tomography:

Positron emission tomography (PET) utilizes positron emitting radioisotopes for scanning. This technique provides valuable information regarding regional tissue biochemistry, physiology, and pharmacology. Gamma-ray emitting radioisotopes of the elements which constitutes the organic substances of the body are taken up for labeling and the selected biochemical or pharmacological compounds. The radionuclides selected for this technique are:

1. Oxygen -15 ($O^{15}$)
2. Nitrogen-13 ($N^{13}$)

3. Carbon-11 ($C^{11}$)

4. Fluorine-18 ($F^{18}$)

These radioisotopes are manufactured by bombarding the stable elements with high-energy ion beams in a cyclotron. During disintegration, they emit a positron and eventually gamma rays by pair annihilation process with electron. Each gamma photon has energy of 511 keV and they travel in opposite directions.

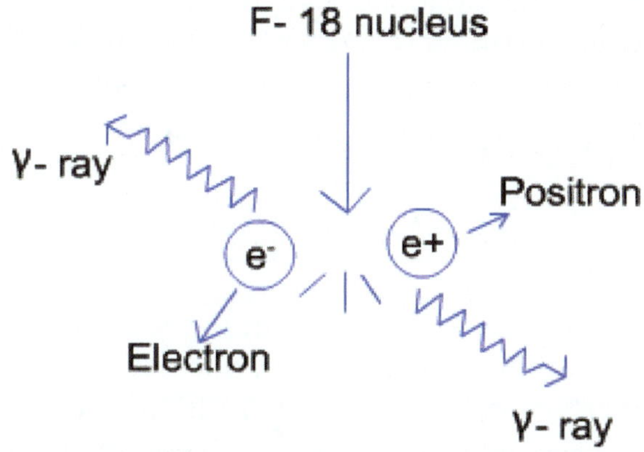


m4.5 Annihilation of electron and positron in PET

Correlated pairs of gamma rays emerge from a body containing positron-emitting radioisotopes. These are detected when they reach a scintillator material in the scanning device, creating a burst of light, which is detected by photomultiplier tubes or silicon avalanche photodiodes (Si APD). The technique depends on simultaneous or coincident detection of the pair of photons. Photons which do not arrive in pairs (i.e., within a few nanoseconds) are ignored.

The most significant fraction of electron-positron decay results in two 511 keV gamma photons being emitted at almost 180 degrees to each other; hence it is possible to localize their source along a straight line of coincidence (also called formally the "line of response" or LOR). In practice the LOR has a finite width as the emitted photons are not exactly 180 degrees apart. If the recovery time of the detectors is in the picosecond range rather than the 10's of nanosecond range, it is possible to calculate the single point on the LOR at which an annihilation event originated, by measuring the "time of flight" of the two photons.

Few applications of PET are given below:

1. Brain utilizes glucose as a primary source of energy, for which detection of regional glucose metabolism using F18 can be successfully done by PET scanning. It is very much sensitive indicator of brain damage and recovery, and has better performance than tissue perfusion studies.

2. PET scanning is very much useful for detection of epileptic patients because the areas show less metabolism and the visualized areas get activated during seizures.

3. PET imaging accurately selects the patients for brain surgery and determines the areas of the brain for surgery.

4. PET is also useful for detection of ischemic areas of myocardium.

## Magnetic Resonance Imaging (MRI)

Magnetic resonance imaging (MRI), formerly referred to as magnetic resonance tomography (MRT), or in chemistry nuclear magnetic resonance (NMR), is a non-invasive method used to render images of the inside of an object. It is primarily used in medical imaging to demonstrate pathological or other physiological alterations of living tissues. The basic physics of MRI is nuclear magnetic resonance (NMR). The basic principle of NMR is the resonance of a radiofrequency signal with the Larmor precessional frequency. The angular momentum vector of a nucleon precesses with a uniform angular frequency around the axis of the magnetic field, in a manner similar to the precession of a mechanical top in a gravitational field. This is known as Larmor precession. Larmor frequency ($\nu_L$) is related to magnetic field (B) by a relation:

$$\nu_L = g\frac{e}{4\pi m}B$$

Where g is the g-factor of the nucleus. It is clear from the relation that vL is independent of L and is a function of charge and mass of the particle and of the magnitude of the applied field B. Clearly, $\nu_L = 0$ if $B = 0$ or $e = 0$. Magnetic moment, associated with the angular momentum also precesses about the field direction. The magnetic field does not change the orientation of the magnetic moment with respect to the field direction. The interaction energy between the magnetic moment μ and the field B is

$$E_{\text{int}} = g\mu_{N.}B$$

where

$$\mu_N = \frac{e\hbar}{2m_p}$$

and

$$g = \frac{\mu}{\mu_N l}$$

$\mu_N$= nuclear magneton,

$m_p$ = proton mass

and

I = total angular momentum

Let us now estimate the Larmor frequency of a proton under an applied field 0.1 W/m²:

$$\nu_{Lp} = g\frac{e}{4\pi m_p}B$$

$$= 5.56 \frac{eB}{4\pi m_p}$$

For

$$B = \frac{1}{10} W / m^2$$

$$v_L = 4 \times 10^6 cycles / \sec$$

This frequency corresponds to the short radiofrequencies. If such a frequency is applied perpendicular to the external field and when the Larmor frequency becomes exactly equal to this frequency the system becomes resonant.

The proton being a fermion, the possible magnetic quantum numbers are +1/2 and -1/2. Also the possible values of the vector components of the magnetic moment in the direction of the applied magnetic field will be $+\mu_B$ and $-\mu_B$. In the absence of the magnetic field, there is no preference for one or other of the two possible magnetic quantum numbers for protons. Thus, in a large sample of protons they are exactly of equal numbers. In the presence of magnetic field, the protons with m = 1/2 will have more tendency to align as compared to the protons with m = -1/2. The population of protons in these two quantum states at equilibrium can be determined by Boltzmann law:

$$\frac{m_{-1/2}}{m_{+1/2}} = e^{-E/kT}$$

where $m_{+1/2}$ and $m_{-1/2}$ are the population of protons in +1/2 and -1/2 states respectively, k is the Boltzmann constant, T is the temperature in absolute scale and E is the interaction energy Eint of the equation. Using the equations, we get

$$\frac{m_{-1/2}}{m_{+1/2}} = e^{-2\mu_r B/kT}$$

$$\sim 1 - \frac{2\mu_p B}{kT}$$

For B = 1/10 W/m², T ~300°K, the fractional difference would be ~ $10^7$. A sample of water of volume ~ 1/10 c.c. contains about 1022 protons. Therefore, in this sample, the excess populations in the lower energy state are ~ 1015 protons. One important parameter that plays role in this is the relaxation time, which is defined as the average lifetime of the excess nuclei in the non-relaxed state. In other words, it is the decay time for reaching thermal equilibrium. We have seen that energy state m = -1/2 being antiparallel to field B, is energetically less favorable than m = +1/2 state. Thus, if an equilibrium distribution (in which both the states have same population) is to be obtained, it is necessary that energy should be lost to the surroundings by the process of nuclear relaxation. Having a knowledge of nuclear magnetic resonance let us ask question:

## What is Magnetic Resonance Imaging (MRI)?

MRI is basically dependent on the relaxation properties of excited hydrogen nuclei. For imaging of any object (selective imaging of different voxels i.e. volume picture elements), orthogonal magnetic gradients are applied. Magnetic gradients are generated by three orthogonal coils, oriented in the x, y and z directions of the scanner. Typical gradient systems are capable of producing gradients from 20 mT/m to 100 mT/m. Basically, the plane of imaging is determined by the field gradients. In a 1.5 T magnetic field (at room temperature), this difference refers to only about one in a million nuclei, since the thermal energy far exceeds the energy difference between the parallel and antiparallel states. Yet the vast quantity of nuclei in a small volume sum to produce a detectable change in field. The bulk collection of nuclei can be partitioned into a set whose sum spin are aligned parallel and a set whose sum spin are anti-parallel. All spatial encoding is obtained by applying magnetic field gradients that encode the position within the phase of the signal. In one dimension, a linear phase with respect to position can be obtained by collecting data in the presence of a magnetic field gradient. In three dimensions (3D), a plane can be defined by "slice selection", in which an RF pulse of defined bandwidth is applied in the presence of a magnetic field gradient in order to reduce spatial encoding to two dimensions (2D). The RF transmission system consists of a RF synthesizer, power amplifier and transmitting coil and is usually built into the body of the MRI scanner.

## How Image Contrast is Achieved in MRI?

The contrast of the MRI image is dependent on two time constants, which are involved in relaxation processes that establish equilibrium following RF excitation. The first one is known as longitudinal relaxation, which is basically relaxation of high-energy nuclei, followed by realignment of nuclear spin with the magnetic field. This is known as 'time 1 or T1' and is typically of the order of 1 sec. The second one is the transverse relaxation, which is related to the local dephasing of spins following the application of the transverse energy pulse. This is known as 'time 2 or T2' and is typically of the order of 100 ms. In order to delineate the areas under interest contrast agents like water, paramagnetic element (gadolinium) or superparamagnetic element (iron oxide nanoparticles) are added. Detailed mathematical analysis for processing the MRI image is based on k-space formalism and is independently formulated by Ljunggren and Tweig.

Detailed discussion on this is out of scope of this book. In clinical practice, MRI is used to distinguish pathologic tissue (such as a brain tumor) from normal tissue. Both CT and MRI scanners can generate multiple two-dimensional cross-sections (slices) of tissue and three-dimensional reconstructions. Unlike CT, which uses only X-ray attenuation to generate image contrast, MRI has a long list of properties that may be used to generate image contrast. By variation of scanning parameters, tissue contrast can be altered and enhanced in various ways to detect different features.

The most important advantage of MRI is that as there is no radioactive isotope present in this, no radiation induced effects/damage takes place during imaging.

## References


- Clarke, R.H.; J. Valentin (2009). "The History of ICRP and the Evolution of its Policies" (PDF). Annals of the ICRP. ICRP Publication 109. 39 (1): 75–110. doi:10.1016/j.icrp.2009.07.009. Retrieved 12 May 2012
- Best, Lara; Rodrigues, George; Velker, Vikram (2013). "1.3". Radiation Oncology Primer and Review. Demos Medical Publishing. ISBN 978-1-62070-004-4

- Bolt, B. A.; Packard, R. E.; Price, P. B. (2007). "John H. Reynolds, Physics: Berkeley". The University of California, Berkeley. Retrieved 2007-10-01
- Litherland, A. E.; Ferguson, A. J. (1961). "Gamma-Ray Angular Correlations from Aligned Nuclei Produced by Nuclear Reactions". Canadian Journal of Physics. 39 (6): 788–824. ISSN 0008-4204. doi:10.1139/p61-089
- Loveland, W.; Morrissey, D.; Seaborg, G.T. (2006). Modern Nuclear Chemistry. Wiley-Interscience. p. 57. ISBN 0-471-11532-0
- Wang, B.; Yan, S.; Limata, B.; et al. (2006). "Change of the 7Be electron capture half-life in metallic environments". The European Physical Journal A. 28 (3): 375–377. Bibcode:2006EPJA...28..375W. ISSN 1434-6001. doi:10.1140/epja/i2006-10068-x
- Martin, B. R. (31 August 2011). Nuclear and particle physics: An introduction (2nd ed.). John Wiley & Sons,. p. 240. ISBN 978-1-1199-6511-4
- The Council of the European Communities (1979-12-21). "Council Directive 80/181/EEC of 20 December 1979 on the approximation of the laws of the Member States relating to Unit of measurement and on the repeal of Directive 71/354/EEC". Retrieved 19 May 2012

7

# Techniques used in Nuclear Science

It becomes difficult for scientists to analyze the composition of materials when the particle's presence is extremely low. This calls for techniques to be used for quantifying and identifying materials. A few of the techniques used in this process are neutron activation analysis, Rutherford backscattering spectrometry, nuclear reaction analysis, etc. The chapter serves as a source to understand the major categories related to nuclear techniques.

## Neutron Activation Analysis (NAA)

Element identification and composition analysis of material is a challenge to scientific community, when it comes down to very low level detection, i.e. when we have to search one unit of some material out of one million unit of another material, which is known as parts per million (ppm). The complexity enhances more when we consider the material even less than ppm level. Similarly, when the question of analyzing very thin layers (nano meter dimension or less) comes, it becomes extremely difficult job. There are several nuclear physics based techniques, such as Rutherford back scattering, nuclear reaction analysis, neutron activation analysis etc. These are extremely sensitive techniques in terms of identification and quantification of materials. One can also analyze very thin l layer of material and with an appropriate analyzing system it is possible to get information of monolayer material deposition.

Conventionally, major, minor and trace element of materials are defined in the following manner:

Major: Concentration exceeding 1% by mass

Minor: Concentration in the range of 0.1% to 1.0 % by mass

Trace: Concentrations less than 0.1% by mass

Different trace levels are

ppm: µg/g (micro gram/gram = 10-6 gm/gm)

ppb: ng/g (nano gram/gram = 10-9 gm/gm)

ppt: pg/g (pico gram/gram = 10-12 gm/gm)

ppf: fg/g (femto gram/gram = 10-15 gm/gm)

ppa: ag/g (atto gram/gram = 10-18 gm/gm)

What is NAA?

In neutron activation analysis (NAA), the sample is exposed to the intense radiation field of a nuclear reactor. The sample is thus bombarded with neutrons, causing the elements to form radioactive isotopes. The radioactive emissions and radioactive decay paths for each element are well- known. Using this information, it is possible to study spectra of the emissions of the radioactive sample, and determine the concentrations of the elements within it. According to the kinetic energies, neutrons can be classified in five groups -- thermal, epithermal, resonance, intermediate and fast. When target nuclei are bombarded with neutrons, the four major reactions that take place are neutron capture, transmutation, fission reaction and inelastic scattering. In neutron capture, the target nucleus absorbs (captures) a neutron, resulting in a product isotope, the mass number of which is incremented by one. If the product nucleus is unstable, it usually de excites by emission of gamma rays and/or β-. For example,

$$^{58}_{26}Fe + ^{1}_{0}n \rightarrow ^{59}_{26}Fe + \gamma$$

In transmutation reaction, the target nucleus absorbs a neutron, emitting charged/non-charged particles like alpha, proton etc. The unstable product nucleus generally de-excites through β- emission back to the target nucleus. Transmutation neutron reactions are caused by neutrons of high energies (fast or intermediate neutrons). Fission reaction is basically fragmentation of heavier nuclei into relatively lower mass nuclei under neutron bombardment. The sequence of events occurring during the most common type of nuclear reaction used for NAA, namely the neutron captures or (n, gamma) reaction is illustrated in the figure. As we know that when a neutron interacts with the target nucleus, the compound nucleus almost instantaneously de-excites into a more stable configuration through emission of one or more characteristic prompt gamma rays. In some cases, there may be emission of delayed gamma rays, but at a much slower rate. This is in accordance with the half-lives of the radioactive nucleus, which could be from a fraction of a second to several years.

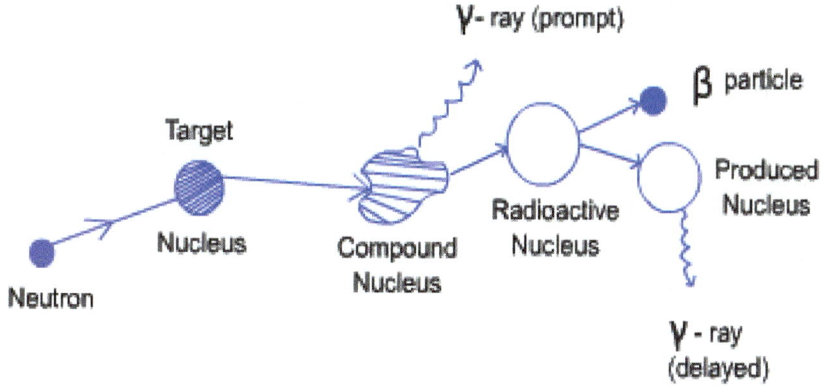


Neutron capture by a target nucleus followed by the emission of gamma rays

## Let us see the Basic NAA Processes

Broadly, there are two types of NAA: prompt-gamma ray NAA or PGNAA and delayed gamma ray NAA or DGNAA. In case of PGNAA, measurements take place during irradiation, and in case of DGNAA, measurements follow radioactive decay. The PGNAA technique is generally performed by using a beam of neutrons extracted through a reactor beam port. It is most applicable to elements (B, Cd, Sm and Gd) with extremely high neutron capture cross-sections. DGNAA is useful for the vast majority of elements that produce radioactive nuclides. The technique is flexible with respect to time, such that the sensitivity for a long-lived radionuclide that suffers from interference by a shorter-lived radionuclide can be improved by waiting for the short-lived radionuclide to decay.

Let us now see the mathematical formulation required to determine the concentration of any element present in any material.

The activity equation for NAA can be obtained in the following manner:1,2

Let,

A = number of decays per second (activity)

N = number of atoms of the target isotope $= \frac{m}{w} \times a \times 6.023 \times 10^{23}$

m = mass of the element in the irradiated sample

a = isotopic abundance

w = atomic weight of the element

λ = decay constant

φ = neutron flux

σ = activation cross section

If $t_{irr}$ is the time of irradiation then we can write

$$A = N\sigma\phi\left[1-\exp\left(-\lambda t_{irr}\right)\right]$$

After a delay of time $t_d$

$$A = N\sigma\phi\left[1-\exp\left(-\lambda t_{irr}\right)\right]\exp\left(-\lambda t_d\right)$$

For a counting time of $t_c$

$$A = N\sigma\phi\left[1-\exp\left(-\lambda t_{irr}\right)\right]\exp\left(-\lambda t_d\right)\left[1-\exp\left(-\lambda t_c\right)\right]$$

Comparator Method:

In this technique, the concentration (at ppm level) of an unknown sample is to compare a standard sample of known concentration. The equation used to calculate the mass of an element of an unknown sample relative to the comparator standard is

$$\frac{A_{sam}}{A_{std}} = \frac{m_{sam}\left(e^{-\lambda T_d}\right)_{sam}}{m_{sam}\left(e^{-\lambda T_d}\right)_{std}}$$

Where A is the activity of the sample (sam) and standard (std), m = mass of the element, λ is the decay constant for the isotope and Td is the decay time. For short irradiation, keeping irradiation, decay and counting times normally same for samples and standards, concentration of the required element in a sample with respect to that of the standard can be written as,

$$C_{sam} = C_{std} \frac{W_{std}}{W_{sam}} \frac{A_{sam}}{A_{std}}$$

where, C is the concentration of any element and W stands for weight of the sample and standard.

## The Non-relative Method

Multi element NAA is possible in non-relative method or single comparator method2. In this technique, standards of all elements are co-irradiated each in turn with the single comparator element. Firstly, the calibration factor for all the elements relative to the comparator element are to be determined and then only the comparator element has to be used in routine measurements instead of individual standards for each element. The calibration factors of course are valid for a specific detector, a specific sample shape, a specific counting geometry and a specific irradiation facility.

## Gamma Ray Detection in NAA

The gamma ray spectrum emitted from an element undergoing activation can consist of gamma ray of few MeV energy, and it may sometimes be extremely complex with some elements having several hundred gamma ray lines. In order to consider the gamma ray detection system to measure gamma rays from neutron activation, the following points should be considered: high detector efficiency, good spectral resolution and count rate. The trade-off between detection efficiency and energy resolution is the main area of uncertainty in the choice of a detection system. The instrumentation used to measure gamma rays from radioactive samples generally consists of a semiconductor, associated electronics and multichannel analyzer. Highly purified germanium (HPGe) detectors, which operate at liquid nitrogen temperature, are also used in many laboratories. Although HPGe detectors come in many different designs and sizes, the most common type of detector is the coaxial detector which are used to measure gamma ray energies over the range from about 60 keV to 3.0 MeV. The schematic of HPGe detector arrangement is given in the figure below:

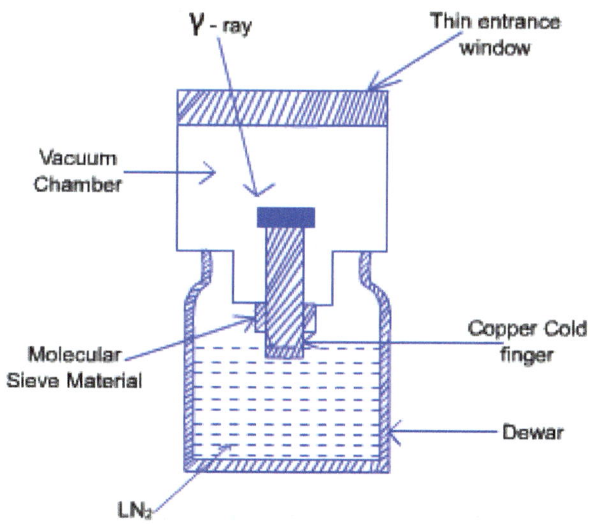


Schematic of HPGe detector arrangement

For bulk material analysis, a high detection efficiency is required and NaI based scintillation detectors are used. Another important aspect in this analysis process is background subtraction.

### Flowchart of NAA Analysis

For NAA gamma ray analysis, the steps which are followed are given below:

Step 1: The gamma ray spectrum of appreciable count is obtained.

Step 2: All peaks are marked as per their energies.

Step 3: The isotope corresponding to the highest peak is identified with the help of standard data table.

Step 4: Next job is to identify whether the isotope can be made by neutron activation using a standard chart3.

Step 5: Once the conditions of making the isotope are applied, the spectrum with the element (look for other intense peaks are matched, then the element has been obtained).

Step 6: If the conditions of making the isotope do not apply, or the spectrum does not match the element, then the next most intense peak has to be identified followed by step 4 and 5. This process has to be repeated until an element that matches the spectrum is obtained.

## Rutherford Backscattering Spectrometry

Rutherford backscattering spectrometry (RBS) is an analytical technique used in materials science. Sometimes referred to as high-energy ion scattering (HEIS) spectrometry, RBS is used to determine the structure and composition of materials by measuring the backscattering of a beam of high energy ions (typically protons or alpha particles) impinging on a sample.

### The Geiger–Marsden Experiment

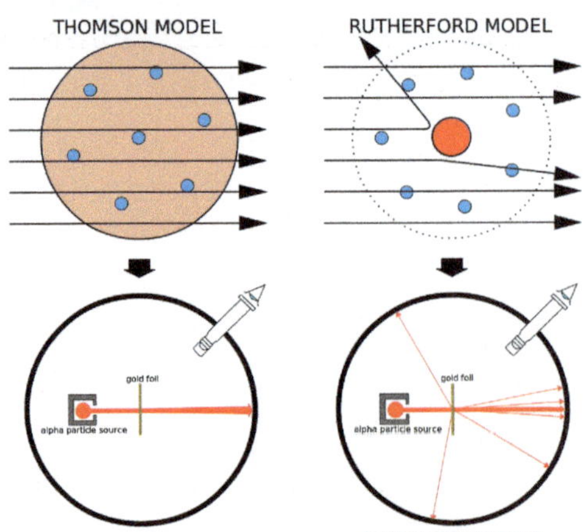


*Left:* Expected results: alpha particles pas ing through the plum pudding model of the atom undisturbed.
*Right:* Observed results: a small portion of the particles were deflected, indicating a small, concentrated positive charge.

Rutherford backscattering spectrometry is named after Lord Ernest Rutherford, a physicist sometimes referred to as the father of nuclear physics. Rutherford supervised a series of experiments carried out by Hans Geiger and Ernest Marsden between 1909 and 1914 studying the scattering of alpha particles through metal foils. While attempting to eliminate "stray particles" they believed to be caused by an imperfection in their alpha source, Rutherford suggested that Marsden attempt to measure backscattering from a gold foil sample. According to the then-dominant plum-pudding model of the atom, in which small negative electrons were spread through a diffuse positive region, backscattering of the high-energy positive alpha particles should have been nonexistent. At most small deflections should occur as the alpha particles passed almost unhindered through the foil. Instead, when Marsden positioned the detector on the same side of the foil as the alpha particle source, he immediately detected a noticeable backscattered signal. According to Rutherford, "It was quite the most incredible event that has ever happened to me in my life. It was almost as incredible as if you fired a 15-inch shell at a piece of tissue paper and it came back and hit you."

Rutherford interpreted the result of the Geiger–Marsden experiment as an indication of a Coulomb collision with a single massive positive particle. This led him to the conclusion that the atom's positive charge could not be diffuse but instead must be concentrated in a single massive core: the atomic nucleus. Calculations indicated that the charge necessary to accomplish this deflection was approximately 100 times the charge of the electron, close to the atomic number of gold. This led to the development of the Rutherford model of the atom in which a positive nucleus made up of *N*e positive particles, or protons, was surrounded by *N* orbiting electrons of charge -e to balance the nuclear charge. This model was eventually superseded by the Bohr atom, incorporating some early results from quantum mechanics.

If the energy of the incident particle is increased sufficiently, the Coulomb barrier is exceeded and the wavefunctions of the incident and struck particles overlap. This may result in nuclear reactions in certain cases, but frequently the interaction remains elastic, although the scattering cross-sections may fluctuate wildly as a function of energy. This case is known as "Elastic (non-Rutherford) Backscattering Spectrometry" (EBS). There has recently been great progress in determining EBS scattering cross-sections, by solving Schrödinger's equation for each interaction.

## Basic Principles

We describe Rutherford backscattering as an elastic, hard-sphere collision between a high kinetic energy particle from the incident beam (the *projectile*) and a stationary particle located in the sample (the *target*). *Elastic* in this context means that no energy is transferred between the incident particle and the stationary particle during the collision, and the state of the stationary particle is not changed. (Except that for a small amount of momentum, which is ignored).

Note that nuclear interactions are generally not elastic, since a collision may result in a nuclear reaction, with the release of considerable quantities of energy. Nuclear reaction analysis (NRA) is useful for detecting light elements. However, this is not Rutherford scattering.

Considering the kinematics of the collision (that is, the conservation of momentum and kinetic energy), the energy $E_1$ of the scattered projectile is reduced from the initial energy $E_0$:

$$E_1 = k \cdot E_0,$$

where k is known as the *kinematical factor*, and

$$k = \left( \frac{m_1 \cos\theta_1 \pm \sqrt{m_2^2 - m_1^2 (\sin\theta_1)^2}}{m_1 + m_2} \right)^2 :$$

where particle 1 is the projectile, particle 2 is the target nucleus, and $\theta_1$ is the scattering angle of the projectile in the laboratory frame of reference (that is, relative to the observer). The plus sign is taken when the mass of the projectile is less than that of the target, otherwise the minus sign is taken.

While this equation correctly determines the energy of the scattered projectile for any particular scattering angle (relative to the observer), it does not describe the probability of observing such an event. For that we need the *differential cross-section* of the backscattering event:

$$\frac{d\omega}{d\Omega} = \left( \frac{Z_1 Z_2 e^2}{4E_0} \right)^2 \frac{1}{(\sin\theta/2)^4},$$

where $Z_1$ and $Z_2$ are the atomic numbers of the incident and target nuclei. This equation is written in the centre of massframe of reference and is therefore not a function of the mass of either the projectile or the target nucleus.

Note that the scattering angle in the laboratory frame of reference $\theta_1$ is *not* the same as the scattering angle in the centre of mass frame of reference $\theta$ (although for RBS experiments they are usually very similar). However, heavy ion projectiles can easily recoil lighter ions which, if the geometry is right, can be ejected from the target and detected. This is the basis of the Elastic Recoil Detection (ERD, with synonyms ERDA, FRS, HFS) technique. RBS often uses a He beam which readily recoils H, so simultaneous RBS/ERD is frequently done to probe the hydrogen isotope content of samples (although H ERD with a He beam above 1 MeV is not Rutherford). For ERD the scattering angle in the lab frame of reference is quite different from that in the centre of mass frame of reference.

Note also that heavy ions cannot *back*scatter from light ones: it is kinematically prohibited. The kinematical factor must remain real, and this limits the permitted scattering angle in the laboratory frame of reference. In ERD it is often convenient to place the recoil detector at recoil angles large enough to prohibit signal from the scattered beam. The scattered ion intensity is always very large compared to the recoil intensity (the Rutherford scattering cross-section formula goes to infinity as the scattering angle goes to zero), and for ERD the scattered beam usually has to be excluded from the measurement somehow.

The singularity in the Rutherford scattering cross-section formula is unphysical of course. If the scattering cross-section is zero it implies that the projectile never comes close to the target, but in this case it also never penetrates the electron cloud surrounding the nucleus either. The pure Coulomb formula for the scattering cross-section shown above must be corrected for this screening effect, which becomes more important as the energy of the projectile decreases (or, equivalently, its mass increases).

While large-angle scattering only occurs for ions which scatter off target nuclei, inelastic small-angle scattering can also occur off the sample electrons. This results in a gradual decrease in ions which penetrate more deeply into the sample, so that backscattering off interior nuclei occurs with a lower "effective" incident energy. The amount by which the ion energy is lowered after passing through a given distance is referred to as the stopping power of the material and is dependent on the electron distribution. This energy loss varies continuously with respect to distance traversed, so that stopping power is expressed as

$$S(E) = -\frac{dE}{dx}.$$

For high energy ions stopping power is usually proportional to $\frac{Z_2}{E}$; however, precise calculation of stopping power is difficult to carry out with any accuracy.

Stopping power (properly, *stopping force*) has units of energy per unit length. It is generally given in thin film units, that is eV/(atom/cm$^2$) since it is measured experimentally on thin films whose thickness is always measured absolutely as mass per unit area, avoiding the problem of determining the density of the material which may vary as a function of thickness. Stopping power is now known for all materials at around 2%.

## Instrumentation

A single stage 2 MeV linear Van de Graaff particle accelerator, here opened for maintenance

An RBS instrument generally includes three essential components:

- An ion source, usually alpha particles ($He^{2+}$ ions) or, less commonly, protons.
- A linear particle accelerator capable of accelerating incident ions to high energies, usually in the range 1-3 MeV.
- A detector capable of measuring the energies of backscattered ions over some range of angles.

Two common source/acceleration arrangements are used in commercial RBS systems, working in either one or two stages. One-stage systems consist of a $He^{+}$ source connected to an acceleration tube with a high positive potential applied to the ion source, and the ground at the end of the acceleration tube. This arrangement is simple and convenient, but it can be difficult to achieve energies of much more than 1 MeV due to the difficulty of applying very high voltages to the system.

Two-stage systems, or "tandem accelerators", start with a source of $He^-$ ions and position the positive terminal at the center of the acceleration tube. A stripper element included in the positive terminal removes electrons from ions which pass through, converting $He^-$ ions to $He^{++}$ ions. The ions thus start out being attracted to the terminal, pass through and become positive, and are repelled until they exit the tube at ground. This arrangement, though more complex, has the advantage of achieving higher accelerations with lower applied voltages: a typical tandem accelerator with an applied voltage of 750 kV can achieve ion energies of over 2 MeV.

Detectors to measure backscattered energy are usually siliconsurface barrier detectors, a very thin layer (100 nm) of P-type silicon on an N-type substrate forming a p-n junction. Ions which reach the detector lose some of their energy to inelastic scattering from the electrons, and some of these electrons gain enough energy to overcome the band gap between the semiconductor valence and conduction bands. This means that each ion incident on the detector will produce some number of electron-hole pairs which is dependent on the energy of the ion. These pairs can be detected by applying a voltage across the detector and measuring the current, providing an effective measurement of the ion energy. The relationship between ion energy and the number of electron-hole pairs produced will be dependent on the detector materials, the type of ion and the efficiency of the current measurement; energy resolution is dependent on thermal fluctuations. It should also be noted that after one ion is incident on the detector, there will be some *dead time* before the electron-hole pairs recombine in which a second incident ion cannot be distinguished from the first.

Angular dependence of detection can be achieved by using a movable detector, or more practically by separating the surface barrier detector into many independent cells which can be measured independently, covering some range of angles around direct (180 degrees) back-scattering. Angular dependence of the incident beam is controlled by using a tiltable sample stage.

## Composition and Depth Measurement

The energy loss of a backscattered ion is dependent on two processes: the energy lost in scattering events with sample nuclei, and the energy lost to small-angle scattering from the sample electrons. The first process is dependent on the scattering cross-section of the nucleus and thus on its mass and atomic number. For a given measurement angle, nuclei of two different elements will therefore scatter incident ions to different degrees and with different energies, producing separate peaks on an N(E) plot of measurement count versus energy. These peaks are characteristic of the elements contained in the material, providing a means of analyzing the composition of a sample by matching scattered energies to known scattering cross-sections. Relative concentrations can be determined by measuring the heights of the peaks.

The second energy loss process, the stopping power of the sample electrons, does not result in large discrete losses such as those produced by nuclear collisions. Instead it creates a gradual energy loss dependent on the electron density and the distance traversed in the sample. This energy loss will lower the measured energy of ions which backscatter from nuclei inside the sample in a continuous manner dependent on the depth of the nuclei. The result is that instead of the sharp backscattered peaks one would expect on an N(E) plot, with the width determined by energy and angular resolution, the peaks observed trail off gradually towards lower energy as the ions pass through the depth occupied by that element. Elements which only appear at some depth inside the sample will also have their peak positions shifted by some amount which represents the distance an ion had to traverse to reach those nuclei.

In practice, then, a compositional depth profile can be determined from an RBS N(E) measurement. The elements contained by a sample can be determined from the positions of peaks in the energy spectrum. Depth can be determined from the width and shifted position of these peaks, and relative concentration from the peak heights. This is especially useful for the analysis of a multilayer sample, for example, or for a sample with a composition which varies more continuously with depth.

This kind of measurement can only be used to determine elemental composition; the chemical structure of the sample cannot be determined from the N(E) profile. However, it is possible to learn something about this through RBS by examining the crystal structure. This kind of spatial information can be investigated by taking advantage of blocking and channeling.

## Structural Measurements: Blocking and Channeling

To fully understand the interaction of an incident beam of nuclei with a crystalline structure, it is necessary to comprehend two more key concepts: *blocking* and *channeling*.

When a beam of ions with parallel trajectories is incident on a target atom, scattering off that atom will prevent collisions in a cone-shaped region "behind" the target relative to the beam. This occurs because the repulsive potential of the target atom bends close ion trajectories away from their original path, and is referred to as *blocking*. The radius of this blocked region, at a distance L from the original atom, is given by

$$R = 2\sqrt{\frac{Z_1 Z_2 e^2 L}{E_0}}$$

When an ion is scattered from deep inside a sample, it can then re-scatter off a second atom, creating a second blocked cone in the direction of the scattered trajectory. This can be detected by carefully varying the detection angle relative to the incident angle.

*Channeling* is observed when the incident beam is aligned with a major symmetry axis of the crystal. Incident nuclei which avoid collisions with surface atoms are excluded from collisions with all atoms deeper in the sample, due to blocking by the first layer of atoms. When the interatomic distance is large compared to the radius of the blocked cone, the incident ions can penetrate many times the interatomic distance without being backscattered. This can result in a drastic reduction of the observed backscattered signal when the incident beam is oriented along one of the symmetry directions, allowing determination of a sample's regular crystal structure. Channeling works best for very small blocking radii, i.e. for high-energy, low-atomic-number incident ions such as $He^+$.

The tolerance for the deviation of the ion beam angle of incidence relative to the symmetry direction depends on the blocking radius, making the allowable deviation angle proportional to

$$\sqrt{\frac{Z_1 Z_2}{E_0 d}}$$

While the intensity of an RBS peak is observed to decrease across most of its width when the beam is channeled, a narrow peak at the high-energy end of larger peak will often be observed, representing surface scattering from the first layer of atoms. The presence of this peak opens the possibility of surface sensitivity for RBS measurements.

## Profiling of Displaced Atoms

In addition, channeling of ions can also be used to analyze a crystalline sample for lattice damage. If atoms within the target are displaced from their crystalline lattice site, this will result in a higher backscattering yield in relation to a perfect crystal. By comparing the spectrum from a sample being analyzed to that from a perfect crystal, and that obtained at a random (non-channeling) orientation (representative of a spectrum from an amorphous sample), it is possible to determine the extent of crystalline damage in terms of a fraction of displaced atoms. Multiplying this fraction by the density of the material when amorphous then also gives an estimate for the concentration of displaced atoms. The energy at which the increased backscattering occurs can also be used to determine the depth at which the displaced atoms are and a defect depth profile can be built up as a result.

## Surface Sensitivity

While RBS is generally used to measure the bulk composition and structure of a sample, it is possible to obtain some information about the structure and composition of the sample surface. When the signal is channeled to remove the bulk signal, careful manipulation of the incident and detection angles can be used to determine the relative positions of the first few layers of atoms, taking advantage of blocking effects.

The surface structure of a sample can be changed from the ideal in a number of ways. The first layer of atoms can change its distance from subsequent layers (relaxation); it can assume a different two-dimensional structure than the bulk (reconstruction); or another material can be adsorbed onto the surface. Each of these cases can be detected by RBS. For example, surface reconstruction can be detected by aligning the beam in such a way that channeling should occur, so that only a surface peak of known intensity should be detected. A higher-than-usual intensity or a wider peak will indicate that the first layers of atoms are failing to block the layers beneath, i.e. that the surface has been reconstructed. Relaxations can be detected by a similar procedure with the sample tilted so the ion beam is incident at an angle selected so that first-layer atoms should block backscattering at a diagonal; that is, from atoms which are below and displaced from the blocking atom. A higher-than-expected backscattered yield will indicate that the first layer has been displaced relative to the second layer, or relaxed. Adsorbate materials will be detected by their different composition, changing the position of the surface peak relative to the expected position.

RBS has also been used to measure processes which affect the surface differently from the bulk by analyzing changes in the channeled surface peak. A well-known example of this is the RBS analysis of the premelting of lead surfaces by Frenken, Maree and van der Veen. In an RBS measurement of the Pb(110) surface, a well-defined surface peak which is stable at low temperatures was found to become wider and more intense as temperature increase past two-thirds of the bulk melting temperature. The peak reached the bulk height and width as temperature reached the melting temperature. This increase in the disorder of the surface, making deeper atoms visible to the incident beam, was interpreted as pre-melting of the surface, and computer simulations of the RBS process produced similar results when compared with theoretical pre-melting predictions.

RBS has also been combined with nuclear microscopy, in which a focused ion beam is scanned across a surface in a manner similar to a scanning electron microscope. The energetic analysis of backscattered signals in this kind of application provides compositional information about the surface, while the microprobe itself can be used to examine features such as periodic surface structures.

# Particle-induced X-ray Emission

Particle-induced X-ray emission or proton-induced X-ray emission (PIXE) is a technique used in the determining of the elemental make-up of a material or sample. When a material is exposed to an ion beam, atomic interactions occur that give off EM radiation of wavelengths in the x-ray part of the electromagnetic spectrum specific to an element. PIXE is a powerful yet non-destructive elemental analysis technique now used routinely by geologists, archaeologists, art conservators and others to help answer questions of provenance, dating and authenticity.

The technique was first proposed in 1970 by Sven Johansson of Lund University, Sweden, and developed over the next few years with his colleagues Roland Akselsson and Thomas B Johansson.

Recent extensions of PIXE using tightly focused beams (down to 1 μm) gives the additional capability of microscopic analysis. This technique, called microPIXE, can be used to determine the distribution of trace elements in a wide range of samples. A related technique, particle-induced gamma-ray emission (PIGE) can be used to detect some light elements.

## Theory

Three types of spectra can be collected from a PIXE experiment:

1. X-ray emission spectrum.
2. Rutherford backscattering spectrum.
3. Proton transmission spectrum.

## X-ray Emission

Quantum theory states that orbiting electrons of an atom must occupy discrete energy levels in order to be stable. Bombardment with ions of sufficient energy (usually MeV protons) produced by an ion accelerator, will cause inner shell ionization of atoms in a specimen. Outer shell electrons drop down to replace inner shell vacancies, however only certain transitions are allowed. X-rays of a characteristic energy of the element are emitted. An energy dispersive detector is used to record and measure these X-rays.

Only elements heavier than fluorine can be detected. The lower detection limit for a PIXE beam is given by the ability of the X-rays to pass through the window between the chamber and the X-ray detector. The upper limit is given by the ionisation cross section, the probability of the K electron shellionisation, this is maximal when the velocity of the proton matches the velocity of the electron (10% of the speed of light), therefore 3 MeV proton beams are optimal.

## Proton Backscattering

Protons can also interact with the nucleus of the atoms in the sample through elastic collisions, Rutherford backscattering, often repelling the proton at angles close to 180 degrees. The backscatter give information on the sample thickness and composition. The bulk sampleproperties allow for the correction of X-ray photon loss within the sample.they are sample but convective process.

## Proton Transmission

The transmission of protons through a sample can also be used to get information about the sample.

## Protein Analysis

Protein analysis using microPIXE allow for the determination of the elemental composition of liquid and crystalline proteins. microPIXE can quantify the metal content of protein molecules with a relative accuracy of between 10% and 20%.

The advantage of microPIXE is that given a protein of known sequence, the X-ray emission from sulfur can be used as an internal standard to calculate the number of metal atoms per protein monomer. Because only relative concentrations are calculated there are only minimal systematic errors, and the results are totally internally consistent.

The relative concentrations of DNA to protein (and metals) can also be measured using the phosphate groups of the bases as an internal calibration.

## Data Analysis

Analysis of the data collected can be performed by the programs Dan32, the front end to gupix.

## Limitations

In order to get a meaningful sulfur signal from the analysis, the buffer should not contain sulfur (i.e. no BES, DDT, HEPES, MES, MOPSO or PIPES compounds). Excessive amounts of chlorine in the buffer should also be avoided, since this will overlap with the sulfur peak; KBr and NaBr are suitable alternatives.

## Advantages

There are many advantages to using a proton beam over an electron beam. There is less crystal charging from Bremsstrahlung radiation, although there is some from the emission of Auger electrons, and there is significantly less than if the primary beam was itself an electron beam.

Because of the higher mass of protons relative to electrons, there is less lateral deflection of the beam; this is important for proton beam writing applications.

### Scanning

Two-dimensional maps of elemental compositions can be generated by scanning the microPIXE beam across the target.

### Cell and Tissue Analysis

Whole cell and tissue analysis is possible using a microPIXE beam, this method is also referred to as nuclear microscopy.

### Artifact Analysis

MicroPIXE is a useful technique for the non-destructive analysis of paintings and antiques. Although it provides only an elemental analysis, it can be used to distinguish and measure layers within the thickness of an artifact.

### Proton Beam Writing

Proton beams can be used for *writing* (proton beam writing) through either the hardening of a polymer (by proton induced cross-linking), or through the degradation of a proton sensitive material. This may have important effects in the field of nanotechnology.

## Nuclear Reaction Analysis (NRA)

In nuclear reaction analysis presence of any element (particularly low mass elements, like C, N, O) is detected through the nuclear reactions emitted almost instantaneously from nuclear reactions produced in the target by the irradiating beam.

One advantage of this analysis is combining it with backscattering spectrometry, highly sensitive depth profiling of elements particularly distributed in the surface and near-surface region of a material can be done. The characteristics of the emitted radiations from various depths depend on the energy loss of the incident ions as they go into the sample and also the charged particles emitted from the reactions as they come out from the sample.

The primary emphasis of this chapter is to discuss how to determine concentration depth profile of various trace element impurities by analyzing some nuclear reactions. First let us see the basic methodologies:

### Basic Methodology

For depth profiling using NRA, two different methods are generally adopted, namely the energy-analysis method and the resonance method.5 In the first case the energy of the analyzing beam is kept fixed and during analysis, the energy spectra of the particles from which the depth profiles are derived, are recorded. The second method known as resonance is used when a sharp peak (resonance) as a function of energy is obtained. In this case, the depth profile is derived from a measurement of the nuclear reaction yield as a function of the energy of the analyzing beam.

## Let us First Discuss the Energy Analysis Method.

In this technique both neutrons and charged particles are used as probes. Let me begin with the following neutron induced reaction 10B(n,α)7Li. The energy of emitted alpha particles is ~1.5 MeV and the reaction cross section is ~4000 barns.6 The charged particles (alpha and Li), which are emitted isotropically due to this reaction start travelling to the outward path and leave the sample and detected in a suitable detection system. The energy of the detected particle depend on the energy loss of the particle in the outward path and the depth (from the surface) where the nuclear reaction takes place. The energy difference (ΔE) of the alpha particle from the surface and from a layer at a depth t is given by:

$$\Delta E = t\left(dE / dx\right)$$

A quantitative analysis of the elemental distribution with depth, i.e. depth profile is generated directly from the charged particle spectrum. The range which can be considered the distance the particles can travel after creation and still exit the surface varies with composition.

Let us now see the nuclear reactions induced by charged particles:

In charged particles induced reactions the incident particle energy is high enough so that it can penetrate through the Coulomb barrier resulting in various nuclear reactions. For example, a deuteron of ~1.7 keV when bombarded on aluminium nitride target results in following nuclear reactions:7

$$^{27}Al\ \left(d,p\right)^{28}Al,\ ^{14}N\left(d,p\right)^{15}N,\ ^{14}N\left(d,\alpha\right)^{12}C.$$

The schematic of the geometry and the reactions associated with this specific example is shown in the figure.

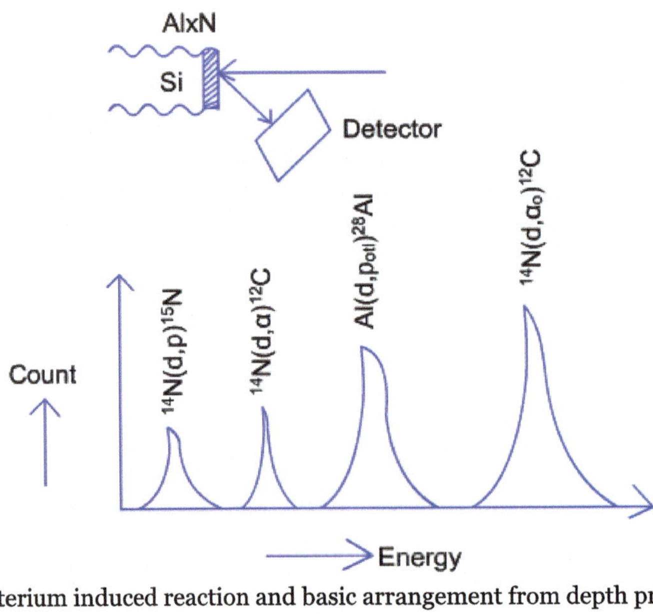


Schematic of deuterium induced reaction and basic arrangement from depth profiling of AlxN/Ni

The elastically scattered particles are stopped in a thin absorber to prevent count rate saturation of the detector and electronic system.

Let us now see the basic equations those govern the layer composition and depth profile using this nuclear reaction process.

Similar to RBS The number of detected particles YD is proportional to the areal concentration (atoms/cm2) Nt and is given by the following equation:

$$Y_D = N_t\ \sigma(\theta) Q\Omega$$

where σ(θ) is the differential cross section, Q is the number of incident particles, and Ω the detection solid angle. However, the cross section in this process cannot be represented by a simple analytical formula. This can be obtained from the nuclear physics literature. For depth profiling, the energy difference between the detected particles originating from the surface and from a depth d depends on the energy losses in inward and outward path.

$$\Delta E = d\left[a\left(dE/dx\right)_{in} + \left(dE/dx\right)_{out}\right]$$

The difference between this and RBS is that the kinematic factor K of RBS is replaced by reaction factor a.6 If the cross section of a reaction is known, the concentration profile can be deduced from the shape of the experimental spectrum.

A resonance peak occurs in many nuclear reactions, which is a function of incident energy. Depth profile is done by measuring the reaction yield as a function of probing ion beam. In practice, the resonances are observed by varying the incident ion energy in small steps and the yield is measured with fluence. Let us first see the following ideal case of resonance as depicted schematically in the figure.

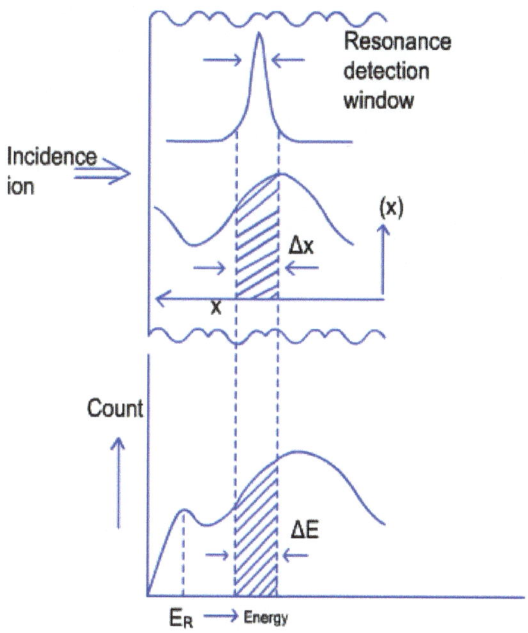


Concentration profile using resonance reaction of NRA

In this case (as seen in lower plot), incident ions of energy E (larger than ER, the resonance energy for reaction with any impurity at trace level) are slowed down until ER is reached at a depth x.

The nuclear reaction occurs at this depth at a rate proportional to the impurity concentration. If the angle between the incident ion and the surface normal of any sample is α, the depth x and the energy of the incident ion can be correlated with the following equation:5

$$E = E_R + (dE/dx)_{in}(x/\cos\alpha)$$

The yield curve as shown in figure can be converted into the desired concentration profile by simply changing scales of yield and energy to corresponding scales of concentration and depth, respectively.

Let us look at the following example:

One classic example to determine depth profile of H implanted in Al2O3 using 19F as a probe beam. It is known that a strong resonance occurs at ~16.4 MeV when a reaction between H and F takes place.

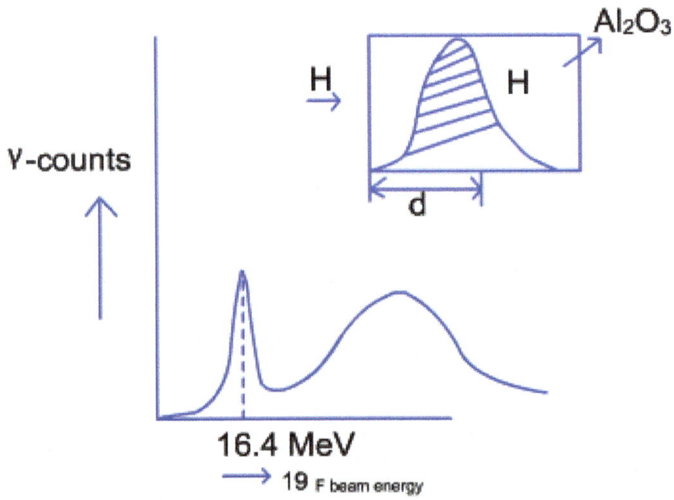


H(19F, αγ) reaction indicating resonance at 16.4 MeV in H implanted Al2O3. (shown in the inset)

The depth profile extracted using above equation is depicted in the figure below considering the reaction yield within the depth d1 to d2.

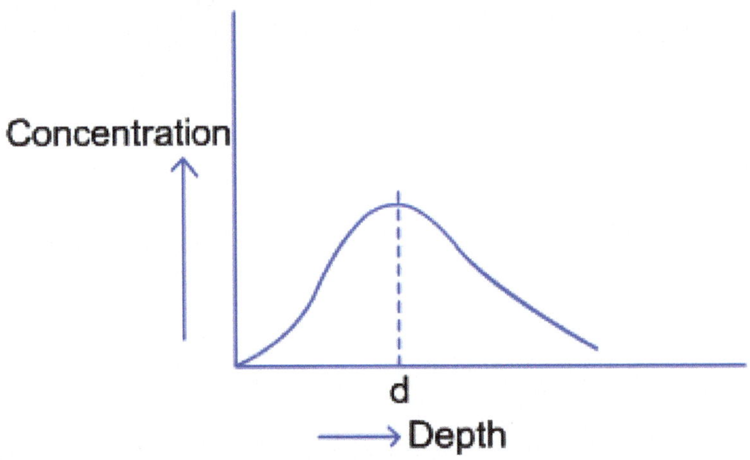


Schematic depth profile of the above system.

Some of the important reactions, which are being routinely used in NRA are listed below:

| Nucleus | Incoming particle | Reaction |
|---|---|---|
| $^{1}H$ | Deuterium (d) | $^{2}H$ (d.p) $^{3}H$ |
| $^{3}He$ | Deuterium (d) | $^{3}He$ (d,p) $^{4}He$ |
| $^{12}C$ | Deuterium (d)- | $^{12}C$ (d,p) $^{13}C$ |
| $^{14}N$ | Deuterium (d) | $^{14}N$ (d, α ) $^{12}C$ |
| $^{16}O$ | Deuterium (d) | $^{16}O$ (d,p) $^{17}O$ |
| $^{7}Li$ | Proton (p) | $^{7}Li$ (p, α ) $^{4}He$ |
| $^{18}O$ | Proton (p) | $^{18}O$ (p, α) $^{15}N$ |
| $^{19}F$ | Proton (p) | $^{19}F$ (p, α ) $^{16}O$ |
| $^{11}B$ | Proton (p) | $^{11}B$(p, α ) $^{8}Be$ |
| $^{31}P$ | Proton (p) | $^{31}P$ (p, α ) $^{28}Si$ |

NRA is used for measuring absolute concentration (atoms/cm2) of light impurities and also provides absolute calibration for other surface sensitive techniques like Secondary ion mass spectrometry, Auger electron spectroscopy etc.

# Permissions

All chapters in this book are published with permission under the Creative Commons Attribution Share Alike License or equivalent. Every chapter published in this book has been scrutinized by our experts. Their significance has been extensively debated. The topics covered herein carry significant information for a comprehensive understanding. They may even be implemented as practical applications or may be referred to as a beginning point for further studies.

We would like to thank the editorial team for lending their expertise to make the book truly unique. They have played a crucial role in the development of this book. Without their invaluable contributions this book wouldn't have been possible. They have made vital efforts to compile up to date information on the varied aspects of this subject to make this book a valuable addition to the collection of many professionals and students.

This book was conceptualized with the vision of imparting up-to-date and integrated information in this field. To ensure the same, a matchless editorial board was set up. Every individual on the board went through rigorous rounds of assessment to prove their worth. After which they invested a large part of their time researching and compiling the most relevant data for our readers.

The editorial board has been involved in producing this book since its inception. They have spent rigorous hours researching and exploring the diverse topics which have resulted in the successful publishing of this book. They have passed on their knowledge of decades through this book. To expedite this challenging task, the publisher supported the team at every step. A small team of assistant editors was also appointed to further simplify the editing procedure and attain best results for the readers.

Apart from the editorial board, the designing team has also invested a significant amount of their time in understanding the subject and creating the most relevant covers. They scrutinized every image to scout for the most suitable representation of the subject and create an appropriate cover for the book.

The publishing team has been an ardent support to the editorial, designing and production team. Their endless efforts to recruit the best for this project, has resulted in the accomplishment of this book. They are a veteran in the field of academics and their pool of knowledge is as vast as their experience in printing. Their expertise and guidance has proved useful at every step. Their uncompromising quality standards have made this book an exceptional effort. Their encouragement from time to time has been an inspiration for everyone.

The publisher and the editorial board hope that this book will prove to be a valuable piece of knowledge for students, practitioners and scholars across the globe.

# Index